MATHEMATICS RESEARCH DEVELOPMENTS

FUZZY LOGIC

APPLICATIONS, SYSTEMS AND TECHNOLOGIES

Mathematics Research Developments

Additional books in this series can be found on Nova's website under the Series tab.

Additional e-books in this series can be found on Nova's website under the e-book tab.

MATHEMATICS RESEARCH DEVELOPMENTS

FUZZY LOGIC

APPLICATIONS, SYSTEMS AND TECHNOLOGIES

DINKO VUKADINOVIC
EDITOR

New York

For permission to use material from this book please contact us:
Telephone 631-231-7269; Fax 631-231-8175
Web Site: http://www.novapublishers.com

NOTICE TO THE READER

LIBRARY OF CONGRESS CATALOGING-IN-PUBLICATION DATA

ISBN: 978-1-62417-151-2

Published by Nova Science Publishers, Inc. † New York

CONTENTS

PREFACE

In the second half of the past century, the fuzzy theory for solving engineering problems with uncertainties has grown enormously. The ability to perform control tasks in unknown environments is one of the most important characteristics of any intelligent control system. In addition, fuzzy logic provides an inference structure that enables the human reasoning capabilities to be applied to artificial knowledge-based systems.

This book presents a few examples of fuzzy logic applications in electrical, chemical, computer and textile engineering with a simple and readable approach. The book is divided into five chapters.

Chapter 1 presents adaptive neuro-fuzzy interface system to predict the pain using the visual scale on patients. The American Hospira company has developed a patient-controlled analgesic device which uses the neuro-fuzzy system to determine the timing of the anesthetic by patient's own discretion and without the need for the medical staff.

In Chapter 2, the authors present application of fuzzy logic techniques to two types of environmental systems. The first type deals with the supervision, diagnosis of acidification states and control of waste treatment plants. Application of fuzzy logic techniques are presented and discussed. In the second type, fuzzy logic techniques are applied to the integration of environmental criteria. The selection of the best packaging materials for a beverage bottle and selection of the best model of children's footwear where used to illustrate this type of the environmental system.

In Chapter 3, two optimization methods have been proposed and applied to solve practical problems related to electric power system operation and control. The first method proposes an adaptation of the gravitational search algorithm to improve the solution of dynamic economic dispatch considering practical generator constraints. The second proposed method based on the particle swarm optimization is proposed for solving multi objective optimal power flow considering distributed shunt STATCOMs. The robustness of the proposed strategy was tested on many standard electrical networks.

In Chapter 4, the authors present the fuzzy logic approach in color matching processes on surface – structural characteristics of textile samples. The main idea was to examine the approach using fuzzy logic based technique in order to provide the method that would include surface parameters of textile samples that are of importance in color applying on structured textile surfaces. It was confirmed that the fuzzy logic reasoning would have its application in control of surface structure influence of colored textiles, especially for lighter shades.

Finally, Chapter 5 presents a few examples of chemical implementation of fuzzy logic operators. Both chromogenic spiro-oxazine and oscillating Belousov-Zhabotinsky reactions allow the fundamental fuzzy logic operators to be implemented. The fuzzy logic systems based on the oscillating Belousov-Zhabotinsky reaction have the advantage of a faster reset time. The fuzzy logic systems based on the chromogenism of spiro-oxazine reaction are particularly appealing because they are based on a defuzzification procedure imitating the way humans distinguish colours.

Dinko Vukadinović

In: Fuzzy Logic
Editor: Dinko Vukadinovic

ISBN: 978-1-62417-151-2

Chapter 1

ADAPTIVE NEURO-FUZZY SYSTEM APPLIED TO PATIENT-CONTROLLED ANALGESIA ANALYSIS

Jiann-Shing Shieh*[*1], Ya-Ting Chan[1], Maysam F. Abbod[2], Wei-Zen Sun[3] and Yeong-Ray Wen[4]

[1]Department of Mechanical Engineering, Yuan Ze University, Taiwan
[2]School of Engineering and Design, Brunel University, London, United Kingdom
[3] Department of Anesthesiology and Center for Emergency Medical Service, College of Medicine, Taipei, Taiwan
[4]Department of Anesthesiology, China Medical University Hospital, Taichung, Taiwan

ABSTRACT

Traditional postoperative pain often needs to rely on nurses regularly applying anesthetics, or when the patient cannot bear to ring the bell for assistance. The waiting time for the arrival of medical personnel is very anguish for the patients and their families, and it may cause patients to raise a question about the hospital. Fortunately, as technology advances, American Hospira developed a patient-controlled analgesia (PCA) device

[*] E-mail: jsshieh@saturn.yzu.edu.tw

to solve the problem. PCA machine enables patients to administer self-anesthetic without medical personnel to achieve the purpose of relieving pain. This chapter uses adaptive neuro-fuzzy inference system (ANFIS) to predict the pain using the visual analogue scale (VAS) on patients. 200 patients' data were used which are divided into 3 parts, training data (60%), validation data (20%) and testing data (20%).

The data is used to train the ANFIS to find the relationship between patient's biological information and the VAS. In order to effectively use these relationships, the ANFIS system rule base has to be simplified by reducing the number of the fuzzy rules.

Statistical methods were utilized to reduce the number of fuzzy rules to the most eight important rules as well as improving the prediction accuracy.

Keywords: Fuzzy neural networks, adaptive neuro-fuzzy inference system, fuzzy rules, patient-controlled analgesia

1. INTRODUCTION

Pain after surgery will cause a huge impact on patient's psychology and physiology. Psychologically, pain not only causes patients discomfort, but also causes the patient's anxiety and fear. Physiologically, pain affects various system dysfunction, including the respiratory system, muscle metabolism, neuroendocrine metabolism etc. Therefore, reducing post-operative pain, not only can solve the patient's emotional problem, but also is a symbol of the hospitals quality of care and medical attention. Moreover, pain is a warning system used to protect the human body, preventing the body from harm and also initiates healing. Nevertheless, too much pain is unnecessary.

If pain is not controlled, the psychological stress and restlessness will make physical condition more deterioration. Therefore, the incidence of pain after surgery is considered to be the most important indicators of analgesic quality in recent years [1]. Postoperative pain often relies on nurses for anesthetics, or rings for the nurse to anesthetize when patient cannot endure the pain. Family members may have question on the hospital for the long time waiting for the medical staff arrival. As technology advances, the American Hospira company developed a patient-controlled analgesic (PCA) device to solve the problem in this regard. PCA can let the patient to determine the timing of the anesthetic by his own without the need for the medical staff.

PCA design concept is based on the patients who assess the pain by himself, and clearly know when their pain must be relieved. The purpose of PCA is to allow patients to evaluate their own demand to control the automatic pump, i.e. trigger a button on the device to activate an injection through microcomputer control.

The PCA device as shown in Figure 1 requires the physician to set each input amount of pain killers, the type of drugs and lock-out interval. Lock-out time is mainly to prevent patients from abuse of painkiller, and it only injected a certain amount of drugs in a certain period of time.

For example, if the lockout time is set to 15 minutes, patients press button three times in this period of time, and it only the first press will give the dosage [2,3]. Visual Analogue Scale (VAS) is an indicator of pain, divided into ten levels in sequence from 0 to 10 [4] with 0 representing not painful, and 10 representing excruciating pain as shown in Figure 2.

In recent years, many scholars use neural network to create the human pharmacokinetic and pharmacodynamic model [5,6,7].

Figure 1. Patient-Controlled Analgesia (PCA).

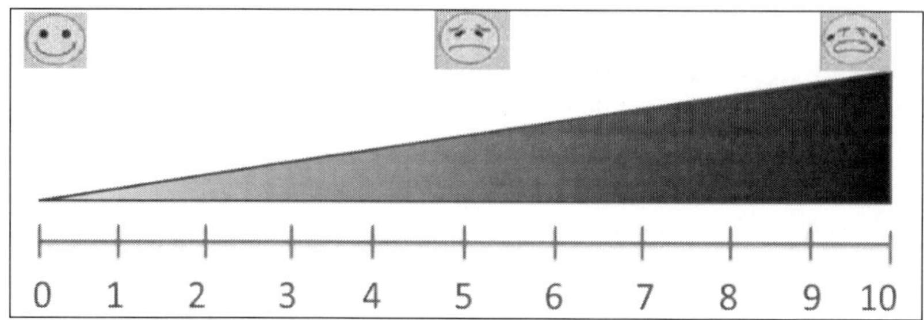

Figure 2. Visual Analogue Scale (VAS).

Some intended to find out the parameters in the pharmacokinetic model, drug dose and time as input to predict the response of human body in different dose to replace the traditional modeling way. However, neural network method still has a shortcoming and it can be presented as a black box, so people cannot understand the contents of the neural network. To overcome this shortcoming, this study hopes to generate comprehensible fuzzy rules by the use of the adaptive neuro-fuzzy inference system.

2. Method

2.1. Neuro-Fuzzy System

The neuro-fuzzy system used in this paper is the adaptive neuro-fuzzy inference the system (ANFIS). ANFIS is a theory based on traditional fuzzy theory, and coupled with neural network as frame [8]. Because of traditional fuzzy inference system lack of the learning mechanism, it must rely on the experts to adjust the membership functions in order to achieve the purposes of reduce errors and increase performance. Artificial neural network (ANN) have a good ability to learn, but it like a black box, and people cannot understand the meaning of inside ANN [9,10]. Therefore, ANFIS can use the learning mechanism of artificial neural network to compensate the unreadable shortcomings of fuzzy inference systems.ANFIS build the fuzzy reasoning ability on neural network, and use the back-propagation error correction rule to adjust the parameters. This network has a total of five layers as shown in Figure 3. This network is composed of fuzzy layer, product layer, normalized layer, defuzzification layer and total output layer [11,12]. Also, it uses fuzzy grid partition method to establish the Sugeno fuzzy inference system [11], uses trapezoidal function as the membership function, and uses the back-propagation learning procedure to construct a group of If-Then rules-based. Then, it gradually adjusts in the appropriate range of membership functions to meet the input-output relationship of fuzzy inference.

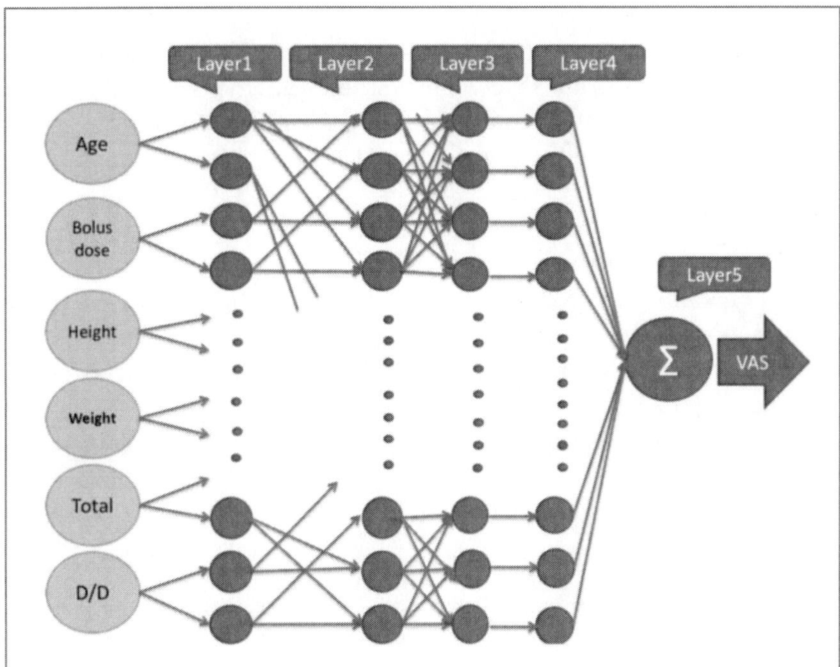

Figure 3. Frame of ANFIS system.

2.2. Parameter Setting of the ANFIS System

2.2.1. Parameter Settings

In this section, the ANFIS system is configured as 6 inputs and one output according to the expert suggestions as shown in Figure 3. The 6 inputs are as follows:

(1) Age of patient
(2) Bolus dose
(3) Height of patient
(4) Weight of patient

(5) Total amount of drug
(6) Delivery / Demand

The 'Bolus dose' means that each time patients press the button, then the amount of drug PCA entered. 'Delivery/Demand' refers to the proportion of the number of PCA actual delivery and total number of patient needs. The output unit is the score of VAS. In this study, the VAS value is selected according to the worst pain score within 24 hours after surgery has finished. The data used in this section consists of 200 patients for general surgery, in which 60% is selected randomly as training data, 20% as validation data and the final 20% as the testing data.

2.2.2. The Overall Structure of System

ANFIS not only can predict patient pain (i.e. VAS) by the learning ability of neural network, but can get fuzzy rules through the fuzzy inference system. This section aims to get the relationship between basic information of the patient and score of VAS.

Using this relationship can provide better care for the patient, as well as a reference to be used by doctors. The database is divided into three parts: training data, validation data and testing data, which follows the procedure shown in Figure 4.

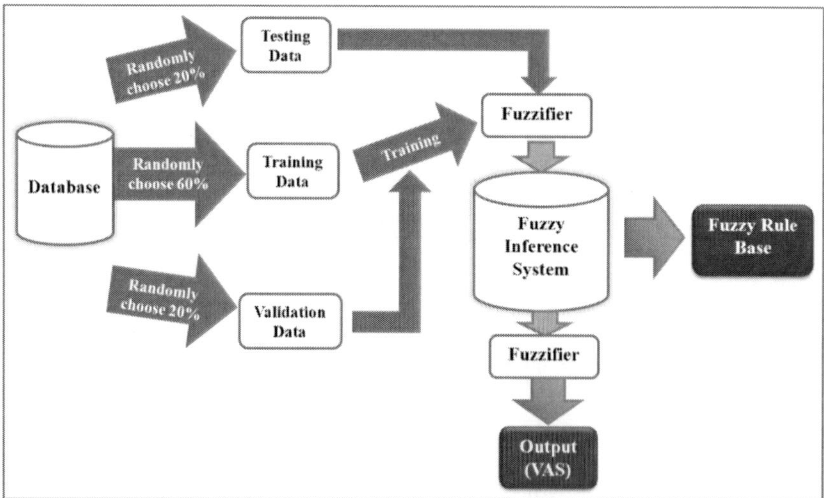

Figure 4. Process frame of ANFIS system.

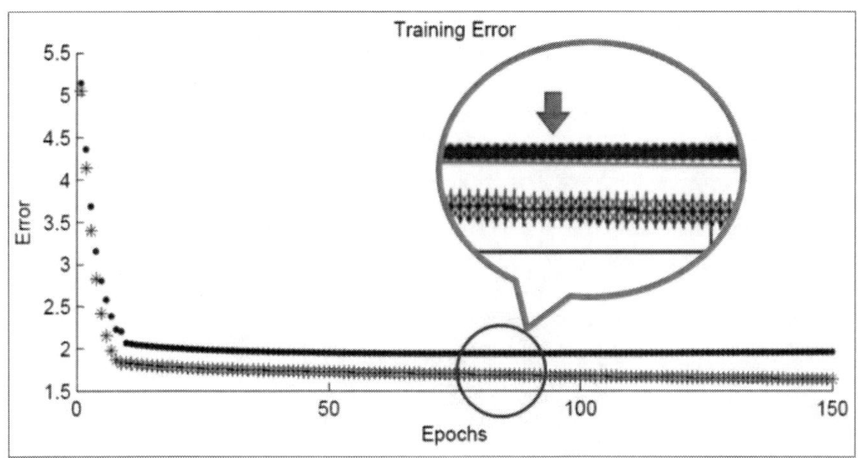

(Note: • is validation error; * is training error. The training error continues to decline but the validation error has started to climb at about 80 epochs. When training is 80 epochs, it must stop in order to avoid over-training).

Figure 5. Training and validation curves.

Using validation data to find the right training times to avoid over training is shown in Figure 5. Testing data do not participate in the training process. It is used into the model to determine the accuracy of the prediction.

3. RESULT

3.1. The Fuzzy Neural Network Model Results

In this section analysis of the most painful results within 24 hours is performed. The data attributes are fed to the neuro-fuzzy network model, whereas the VAS is considered as the output. The predicted results are compared the actual value of VAS.

The membership function for each variable is divided into two levels (high and low) for convenient interpretation of fuzzy rules, and ANFIS system has generated a total of 64 rules.

The network architecture is shown in Figure 6. After training for 80 epochs, the average root mean square error (RMSE) of testing data is 2.3625, as shown in Figure 7.

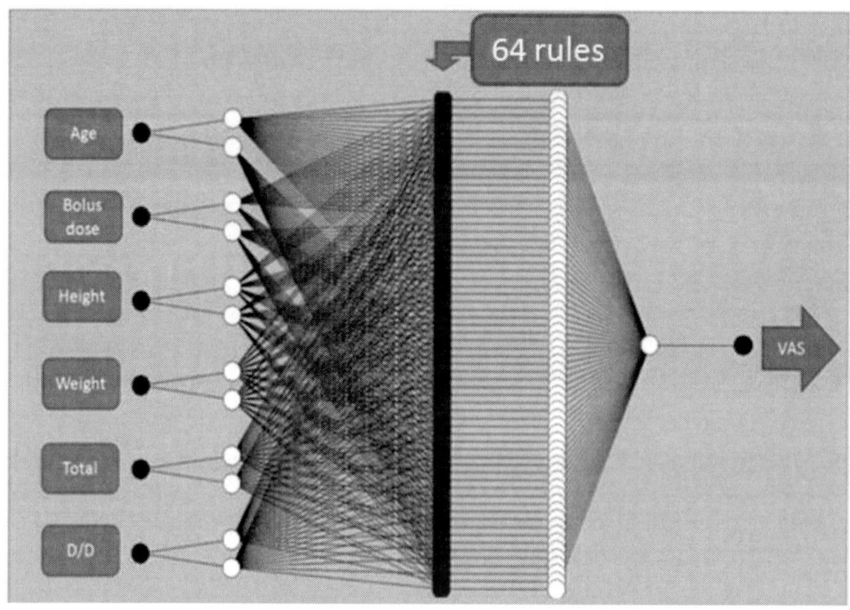

Figure 6. ANFIS architecture, including a total of $2^6 = 64$ fuzzy rules.

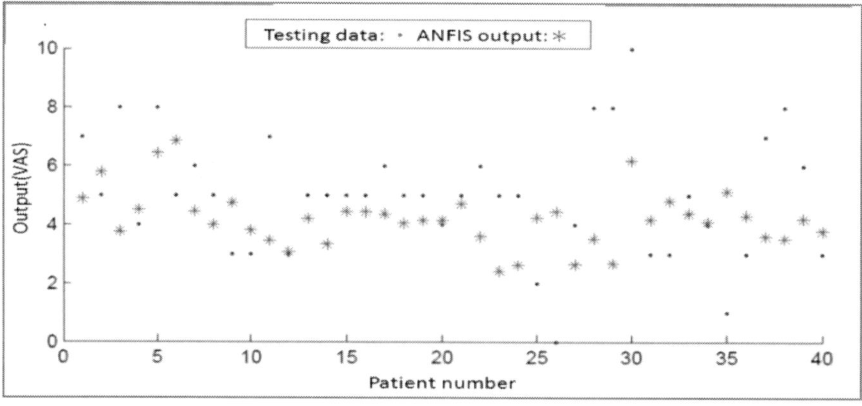

Figure 7. VAS within 24 hour testing error (64 rules), the average error is 2.3625.

3.2. Fuzzy Rule Base Reduction

Given too many rules will make it difficult for medical doctors to interpret the rules. Therefore, extracting the important rules from the original 64 rules is a feasible approach which makes the rules easy to read. In order to achieved this task, the original training data and validation data into the model, and counts the more frequent fired rules before defuzzification of the 64 rules as shown in Figure 8. Statistical results are displayed in Figure 9. In this study, the rules that are fired over 20 times, 30 times, 40 times and 50 times are considered and this is performed for the testing data. In order to decide the optimum number of rules, the relationship between the number of rules and output errors is observed.

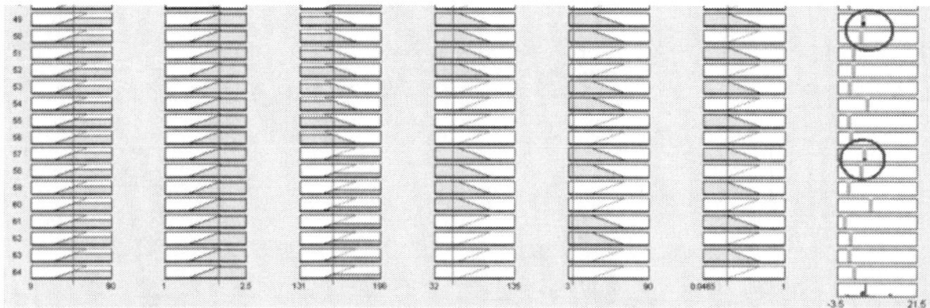

(Note: The yellow part represents how fit with each input for each rule; the dark blue section which mark red circle represents the weight of output value. Regardless of the amount of weight, as long as show the dark blue part, it counts as one).

Figure 8. The weights of fuzzy rules.

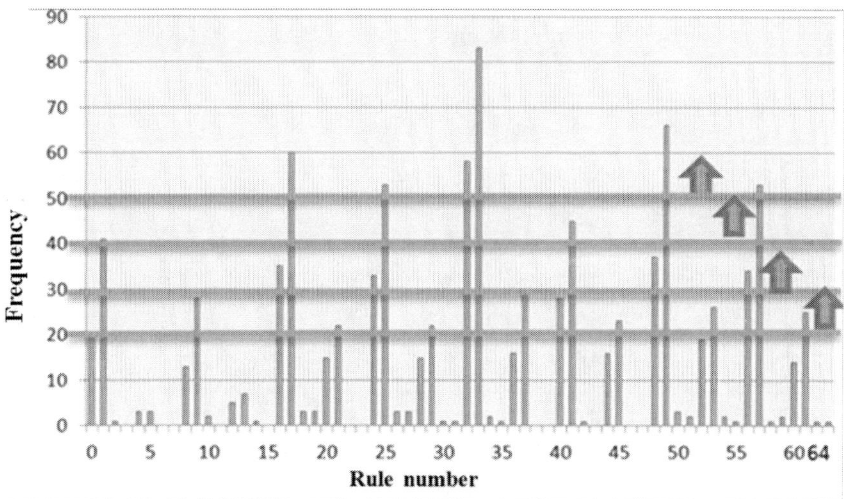

Figure 9. Statistical rules firing results.

According to the statistical results, there are 21 rules fired over 20 times which the average RMSE of the testing data is 2.3932. There are 13 rules fired over 30 times which the average RMSE of the testing data is 2.32609. There are 8 rules fired over 40 times which the average RMSE of the testing data is 2.1831. There are 6 rules fired over 50 times which the average RMSE of the testing data is 2.1913. As shown from Figures 10 to 13, Red (*) denotes the VAS values from ANFIS predictions, and Blue (.) denotes the actual VAS values. For convenience, the results are summarized as shown in Table 1 and graphically presented as a chart as shown in Figure 14.

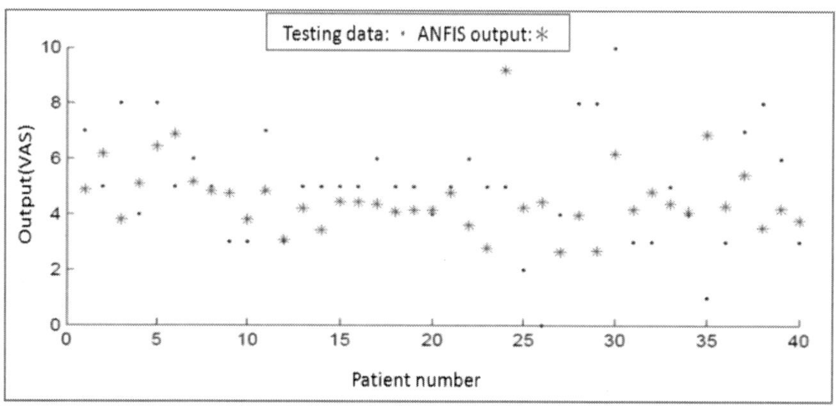

Figure 10. VAS within 24 hour testing error (21 rules), the average RMSE is 2.3932.

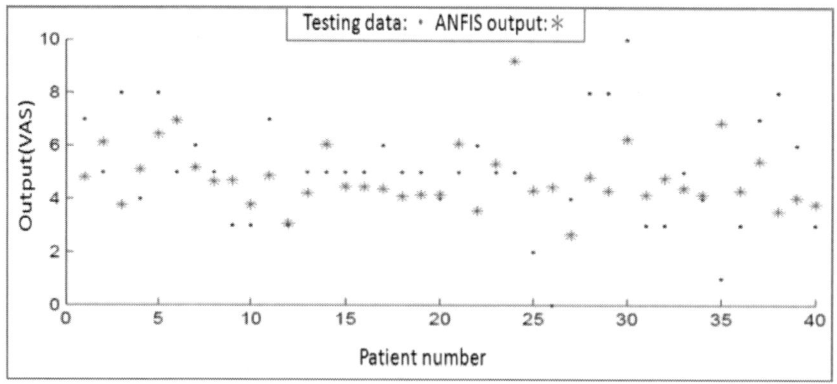

Figure 11. VAS within 24 hour testing error (13 rules), the average RMSE is 2.2609.

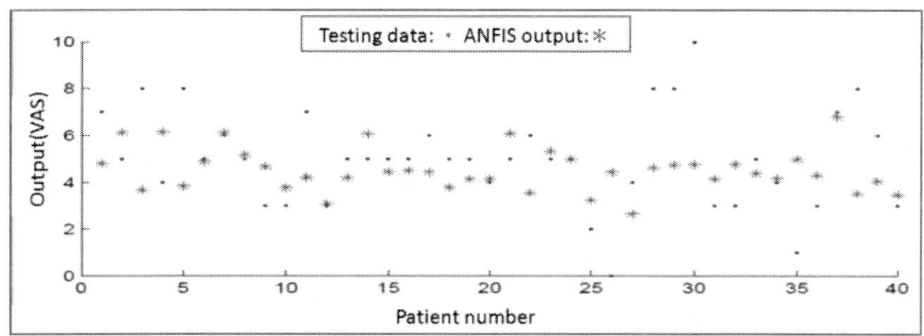

Figure 12. VAS within 24 hour testing error (8 rules), the average RMSE is 2.1831.

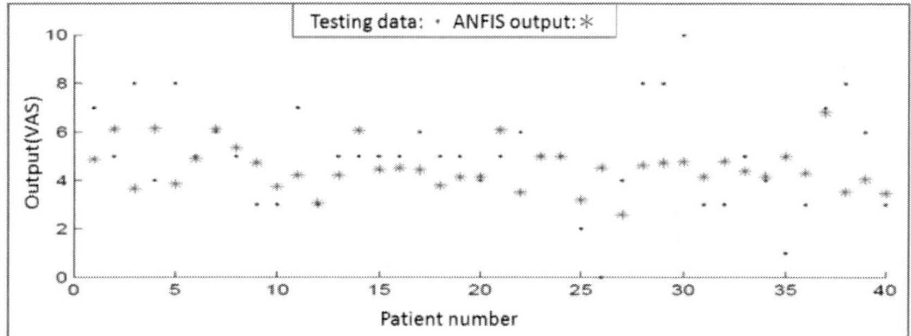

Figure 13. VAS within 24 hour testing error (6 rules), the average RMSE is 2.1913.

The model which has more rules does not mean its predicting ability will better than having fewer rules. According to the results, it is speculated that the number of rules does not have a certain relationship with testing error, but according to different models, the best number of rules will be different.

Table 1. RMSE of testing error for different number of rules

Firing Frequency	Average Testing Error	Rule No.
> 0	2.3625	64
> 20	2.3932	21
> 30	2.2609	13
> 40	2.1831	8
> 50	2.1913	6

The model in this study shows that when it is reduced to only 8 rules, the predicted value of VAS is closest to the values of the original VAS. It means that reducing the rules from 64 to 8, the model's prediction accuracy is better than the original model. Therefore, the existing ANFIS architecture can greatly be simplified as shown in Figure 15.

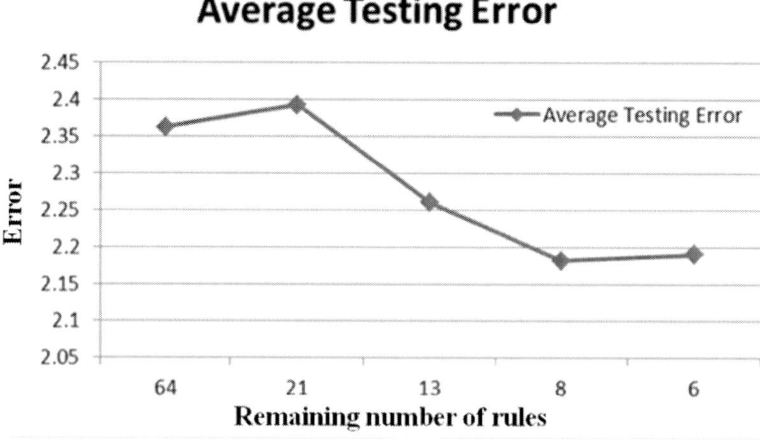

Figure 14. Comparison of the average testing error of RMSE and the remaining number of rules.

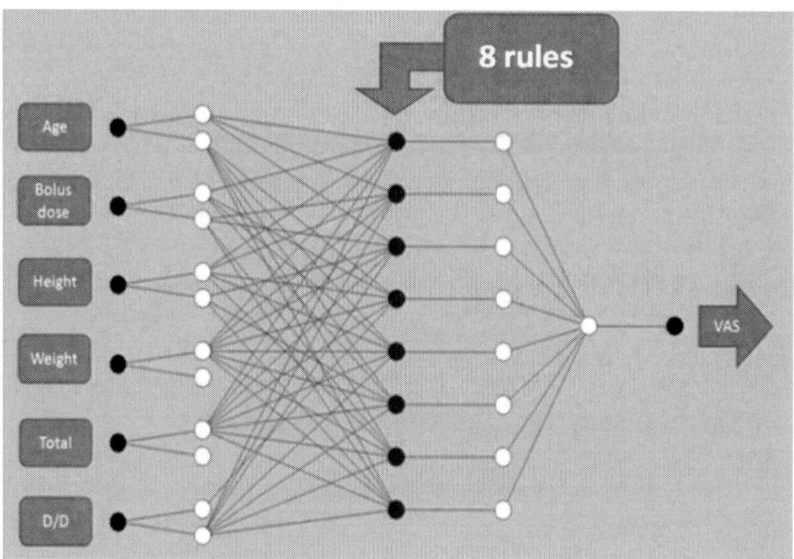

Figure 15. ANFIS architecture (8 fuzzy rules).

3.3. Linguistic Rules Table

In order to make the fuzzy rules easier to interpret for the application, the fuzzy rules are extracted from ANFIS model so that the network can be understood [13]. From the result, it is clear that the model has 8 fuzzy rules which can predict the closer VAS values. Therefore, the 8 rules that are included in the linguistic rules base is listed as shown in Table 2.

Table 2. Linguistic rules table

	Age	Bolus dose	Height	Weight	Total	D/D		VAS
IF	LOW	LOW	LOW	LOW	LOW	HIGH	THEN	MEDIUM
IF	LOW	HIGH	LOW	LOW	LOW	HIGH	THEN	LOW
IF	LOW	HIGH	HIGH	LOW	LOW	HIGH	THEN	MEDIUM
IF	HIGH	LOW	LOW	LOW	LOW	LOW	THEN	MEDIUM
IF	HIGH	LOW	LOW	LOW	LOW	HIGH	THEN	MEDIUM
IF	HIGH	LOW	HIGH	LOW	LOW	HIGH	THEN	MEDIUM
IF	HIGH	HIGH	LOW	LOW	LOW	HIGH	THEN	MEDIUM
IF	HIGH	HIGH	HIGH	LOW	LOW	HIGH	THEN	MEDIUM

(Note: the age set 10-year-old to 50 years old as low, 50 to 90 years as high; Bolus dose set 1 ~ 1.75mg as low, 1.75 ~ 2.5mg as high; height set the boundaries for the 165cm; weight set the boundaries for the 82kg; total dose set the boundaries for the 49mg; the Delivery / Demand ratio set the boundaries for 0.52. VAS is divided into three levels, 0 to 4 (not painful), 4 to 7 (painful), 7 to 10 (very painful).

4. Discussion

Architecture of ANFIS was utilized in this study with two membership function per variable, only high and low, and constructed $2^6 = 64$ rules. Furthermore, when the input membership function are increased to three labels: high, medium, and low (there will be $3^6 = 729$ rules), the large number of rules may not give the best prediction results. In addition, patient data only randomly divide into testing data, validation data and training data without cross validation. In the future, cross-validation can be considered in order to makes the results more accurate. There is a very interesting point that can be observed from Table 2: the most important rules that have been inferred do not have the consequent part of "very painful" fuzzy label. The reason for this might be due to the fact that patients after surgery with the help of pain killers have not reached to the very painful situation. To know what kind of circumstances that will cause the patient to feel severe pain is still very important, so this problem can be solved according to the experience of experts. The experts can provide one or two rules that can be appended to the rule base so that a complete linguistic rules base can be realized.

CONCLUSION

This chapter utilizes the adaptive neuro-fuzzy inference system for modeling pain system, and with simple statistical method the rule base is simplified to the most 8 important rules from large number of fuzzy rules. In comparison with neural network, fuzzy rules are easier to understand. The difficulty of this application is the dramatic increase in the number of rules when the fuzzy partition is increased. In order to solve this shortcoming, this study has established a simplified method to get the most important fuzzy rules for the adaptive neuro-fuzzy inference system. A general methodology has been established for those who require simplifying the fuzzy rules.

REFERENCES

[1] E. Idvall, "Post-Operative Patients in Severe Pain but Satisfied with Pain Relief," *Journal of Clinical Nursing*, 11(6), 841-842, 2002.

[2] P.H. Sechzer, "Studies in pain with the analgesic-demand system", *Anesthesia and analgesia*, vol.50, pp.1-10, 1971.

[3] Y.K. Chang, "A Hierarchical i-Pain System Applied in Patient-Controlled Analgesia Analysis." *Master Thesis in Yuan Ze University,* 2006.

[4] W.C. Chong, "Ensembled Neural Networks Applied in Pain Evaluation of Orthopedics Post-operation via Patient Controlled Analgesia", Bachelors Thesis, Yuan Ze University, 2010.

[5] P. Veng-Pedersen, N.B. Modi, "Application of Neural Networks to Pharmacodynamics," *Journal of Pharmaceutical Sciences*, Vol. 82, No. 9, pp.918-926, 1993.

[6] J.V.S. Gobburu, E.P. Chen, "Artificial Neural Networks as a Novel Approach to Integrated Pharmacokinetic-Pharmacodynamic Analysis," *Journal of Pharmaceutical Sciences*, Vol. 35.No. 5, pp. 505-510, 1996.

[7] J. Opara, S. Primozic, P. Cvelbar, "Prediction of Pharmacokinetic Parameters and the Assessment of Their Variability in Bioequivalence Studies by Artificial Neural Networks", *Pharmaceutical Research*, Vol. 16, No. 6, pp. 944-948, 1999.

[8] C.T. Lin, C.S.G. Lee, "*Neural Fuzzy System A Neuron-Fuzzy Synergism to Intelligent Systems*," Gau-Lih Publisher, 1999.

[9] C. H. Wei, I. C. Chen, "Review of Artificial Neural Network Research and Applications in Transportation", *Transportation Planning Journal*, Vol. 30, No. 2, June 2001.

[10] S.R. Huanga, C.Y. Linc, C.C. Wub, S.J. Yanga, "The application of Fuzzy controller for fuel cell generating studies", *International Journal of Hydrogen Energy*, Vol. 3, No. 3, 2008.

[11] A.Z. Cao, L. Tian, "Application of the Adaptive Fuzzy Neural Network to Industrial Water Consumption Prediction", *Journal of Anhui University of Technology and Science*, Vol. 20, No. 3, Sep. 2005.

[12] H.C. Su, "*Using Based fuzzy Inference System to Improve the Network Prediction of Effluent Quality from Wastewater Treatment Plant*", Chaoyang University of Technology, 2008.

[13] J.W.F. Catto, M.F. Abbod, D.A. Linkens and F. C. Hamdy, "Neuro-Fuzzy Modeling: An Accurate and Interpretable Method for Predicting Bladder Cancer Progression" *The Journal of Urology*, Vol. 175, pp. 474-479, February 2006.

In: Fuzzy Logic
Editor: Dinko Vukadinovic

ISBN: 978-1-62417-151-2

Chapter 2

APPLICATION OF FUZZY LOGIC REASONING IN ENVIRONMENTAL SYSTEMS

Eugenio F. Carrasco, Carlos García-Diéguez, Marta Herva and Enrique Roca*

Sustainable Processes and Products Engineering and Management Group
Department of Chemical Engineering, School of Engineering
University of Santiago de Compostela, Santiago de Compostela, Spain

ABSTRACT

In many cases, the solution for the design and control of environmental systems requires processing imprecise, uncertain, qualitative and vague data. Two cases of this type of processes are the supervision and control of wastewater treatment plants, specifically for anaerobic treatment processes, and the ecodesign of processes and products, which requires considering environmental criteria apart from the general design criteria (i.e. technical, functional, ergonomic, aesthetic or economical). In this chapter, the application of fuzzy logic techniques to these environmental systems is presented.

The first system deals with the supervision, diagnosis of acidification states and control of wastewater treatment plants (for wastewaters containing ethanol and carbohydrates). Representative experiments of

* E-mail: enrique.roca@usc.es.

application of fuzzy logic techniques are presented and discussed for these three items.

In the second one, fuzzy logic techniques are applied to the integration of environmental criteria in ecodesign. To do this, a number of environmental evaluation methodologies (Ecological Footprint, Life Cycle Assessment, Environmental Risk Assessment, etc.) are considered to build up an environmental index, which takes into account the values of different indicators at a time. The integration of all of them was made using fuzzy logic reasoning. As a result, a Fuzzy Ecodesign Index was obtained. A case study consisting in the selection of the best packaging materials for a beverage bottle is presented to illustrate this application.

1. Introduction

Fuzzy logic (FL) is one of the most common methodologies used to address uncertainty matters (Bellman and Zadeh, 1970). The use of FL techniques allows obtaining a quantitative approach using a qualitative representation (Zadeh, 1965). The most important features of FL are that it uses linguistic variables and that knowledge is represented by if-then linguistic rules. A linguistic variable is defined by four items: (1) the name of the variable; (2) its linguistic values; (3) the membership functions of the linguistic values; (4) the physical domain over which the variable takes its quantitative values (Phillis and Andriantiatsaholiniaina, 2001).

FL presents flexibility and tolerance with imprecise data. It can be built on top of human experience, combining natural language in an easy-to-understand way and also being able to model complex non-linear functions. All these reasons, together with the possibility of combination with conventional control techniques, make FL a very good tool for using with environmental systems.

FL techniques have been applied in a number of studies in the environmental field. The first environmental field considered in this chapter is the anaerobic treatment of wastewater, which involves processes that are carried out by a heterogeneous population of micro-organisms (Lettinga, 1995). These processes are performed in bioreactors where many different transport phenomena and complex flow distributions take place. These factors make quite difficult to achieve good process representations using mathematical models, in spite of some efforts dealing with the development of models for anaerobic digestion (Bernard *et al.*, 2001; Siegrist *et al.*, 2002; Batstone *et al.*, 2002). In view of this, advanced knowledge-based control systems appear as an adequate alternative. The introduction of expertise in a control system can be made by means of a rule base constituting an expert system. The expert system must be able to supply a reliable diagnosis based on the evolution of the variables of the process, giving the possible solutions to restore the stable operation of the reactor, when a failure takes place (Pullammanappallil *et al.*, 1998). Several authors (Müller *et al.*, 1997; Giraldo-Gómez and Duque, 1998; Steyer *et al.*, 1999) implemented fuzzy-logic-based expert systems in order to establish an automatic advanced control for anaerobic digesters of wastewater treatment plants.

The second environmental field considered in this chapter is the ecodesign of products by defining an ecodesign index that integrates the criteria provided by three complementary environmental evaluation methodologies (Ecological Footprint, Life Cycle Assessment and Environmental Risk Assessment). Several authors studied the application of FL to this environmental field. Ocampo-Duque *et al.* (2006) developed a Fuzzy Water Quality (FWQ) index calculated with fuzzy reasoning and using Analytical Hierarchy Process (AHP) for weight assignment to the variables involved in the rules, thus defining the relative importance and influence of the input parameters in the final score. This final index integrated a wide set of indicators including organic pollution, nutrients, pathogens, physicochemical macro-variables, and priority micro-contaminants, and was intended to assess water quality in rivers. Similarly, Sadiq and Husain (2005) proposed the use of a fuzzy-based methodology and a three-stage hierarchical structure for estimating aggregative risk of various environmental activities, pollution sources and routes in a given process. The developed methodology was applied to a case study of offshore drilling waste for evaluating various discharge scenarios. Phillis and Andriantiatsaholiniaina (2001) developed a model called Sustainability Assessment by Fuzzy Evaluation (SAFE) in which ecological and human inputs were treated individually and then combined with the aid of FL to provide an overall measure of sustainability (Andriantiatsaholiniaina *et al.*, 2004). Marchini *et al.* (2009) presented F-IND, a user-friendly software framework for the development of multivariable indices with a fuzzy approach. The framework was found suitable by the time the work was carried out for any kind of ecological system (e.g. air, soil, water) and could be used for evaluation of quality, vulnerability, sustainability, impact magnitude or any other ecosystem property. This allowed environmental technicians not familiar with the theoretical fundamentals of FL to convert their expert knowledge into the desired fuzzy index in a simple way. Finally, the fuzzy multi-objective model proposed by Kuo *et al.* (2009) aimed at considering not only environmental criteria through a Life Cycle Assessment (LCA), but also the customer needs and cost considerations in the ecodesign of products. They developed an eco-quality function deployment (Eco-QFD) to aid product design teams in seeking the overall customer satisfaction.

2. Case Studies

2.1. Case Study I: Anaerobic Wastewater Treatment Plant

This first case study applies FL to the determination of acidification states and control of an anaerobic wastewater treatment plant (Carrasco *et al.*, 2002, 2004; García *et al.*, 2007). The diagnosis system uses experts' knowledge to determine the acidification state of the process. The architecture of the monitoring and control system can be seen in Figure 1. A PLC is used as an interface between the wastewater treatment plant (WWTP) and the computer, where the expert system is implemented. Initially, only wastewaters containing ethanol as organic source were employed.

A schematic layout of the process plant is shown in Figure 2. Wastewater is collected in a feeding tank for homogenization. Then, it is pumped to a settler where solids are removed with the help of a flocculant solution. This pretreatment allows the separation of an important fraction of recalcitrant compounds (40% of the lignin and 80% of the phenols) within the eliminated solids (about 70%), minimizing their possible toxic effect on the biomass. A hybrid anaerobic digester (UASB-AF) with a volume of 1.1 m^3 was used for the biological treatment. The bioreactor is equipped with a biogas flow-meter, feed and recycling flow-meters, two thermometers Pt-100, a biogas (CH_4 and CO) analyzer, a hydrogen analyzer and a standard pH-meter. All data were measured continuously by the sensors and recorded every 15 min. This sampling time represents 1.04% of the hydraulic retention time of the bioreactor.

In order to develop the diagnosis system, the Matlab Fuzzy Logic Toolbox was used. In the first step, input variables were selected from which the information about the state of the process would be determined. The selected variables were: biogas flow rate (GF), percentage of methane in the biogas (CH_4), carbon monoxide concentration in the biogas (CO), feeding flow rate (FF) and pH.

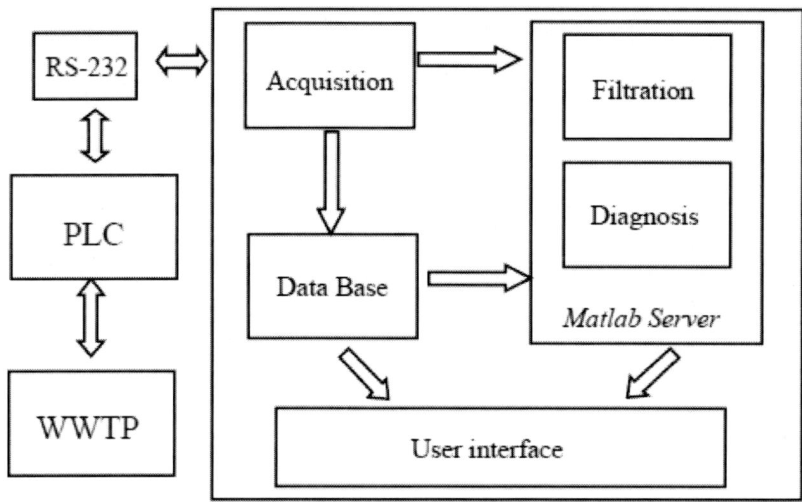

Figure 1. Architecture of the system.

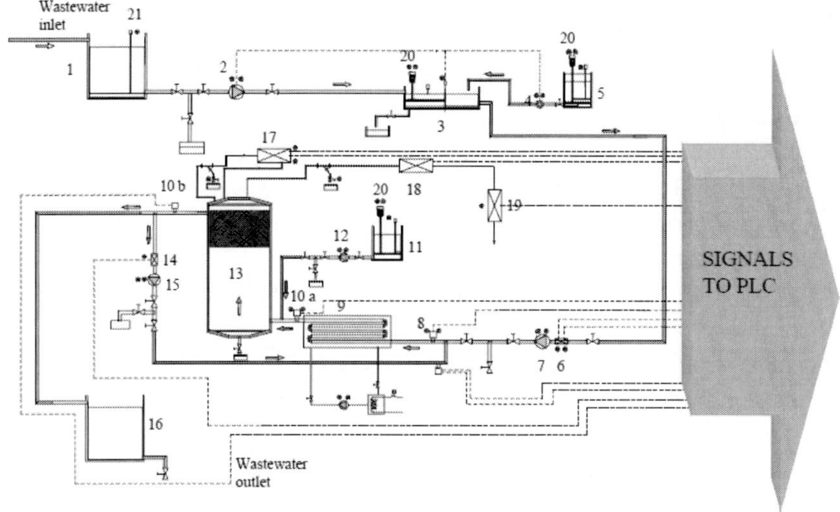

Figure 2. Schematic layout of the plant: (1) Feeding tank; (2) feeding pump #1; (3) settling tank; (4) flocculant addition pump; (5) flocculant storage tank; (6, 14) liquid flow-meters; (7) feeding pump #2; (8) pH-meter; (9) heat exchanger; (10a, b) Pt-100; (11) nutrients tank; (12) nutrients pump; (13) hybrid digester; (15) recycling pump; (16) effluent tank; (17) gas analyzer (CO and CH_4); (18) gas flow-meter; (19) H_2 analyzer; (20) stirrer; (21) level sensor.

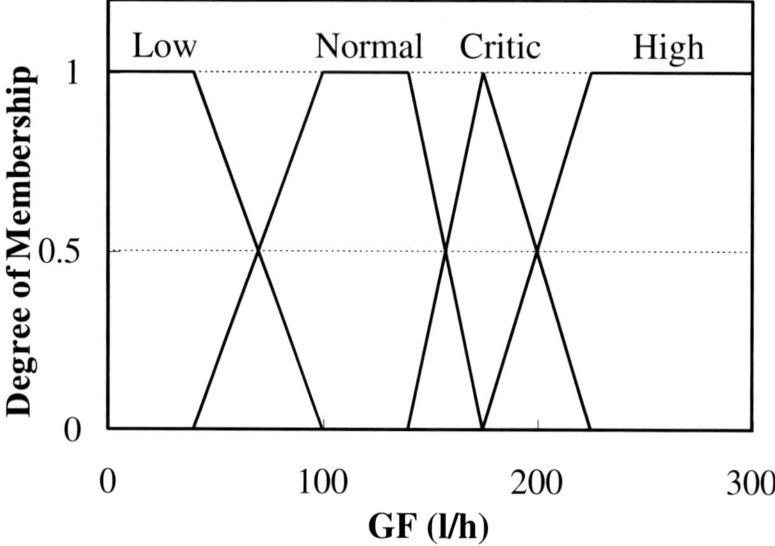

Figure 3. Membership functions for biogas production.

The membership functions were determined using expert knowledge, considering the process and the water to be treated. Figure 3 illustrates the membership functions for the biogas production for the anaerobic treatment of this wastewater from the fiber-board factory. The shape of these functions depends on the process and on the wastewater and may differ quantitatively more than qualitatively with other wastewaters.

The next step was the determination of the possible values for the output, called state, of the diagnosis system. Seven possible results were taken as fuzzy sets with their corresponding membership functions. Hence, each possible result corresponds to a number: Organic Overload, meaning High Acidification by Organic Overload (1), Medium Acidification by Organic Overload (2), Low Acidification by Organic Overload (3), Normal (4), Low Acidification by Hydraulic Overload (5), Medium Acidification by Hydraulic Overload (6) and Hydraulic Overload or high acidification by hydraulic overload (7). These membership functions are only singleton spikes.

Taking into account the inputs and the possible diagnosis results, a structure of knowledge-based rules, a total of 132 rules were defined and implemented, the following being a typical example of one of them:

IF GF is *low* and CH_4 is *very low* and FF is *normal*, **THEN** state is *organic overload*

As a representative result, a typical situation of progressive acidification due to a hydraulic overload is presented. In this experiment, a more concentrated wastewater was employed, namely with a COD concentration of 15 g/l, and the reactor was operated under high hydraulic and organic loads. An organic loading rate (OLR) of 15 kg COD/m^3 d and a HRT of 1 d were used. Once the results about the acidification state of the process are obtained, they are displayed to the operator, who must decide the best actions to carry out. However, diagnosis results could be input into another fuzzy supervision system able to estimate the best actions and even close the control loop (Puñal *et al.*, 2001).

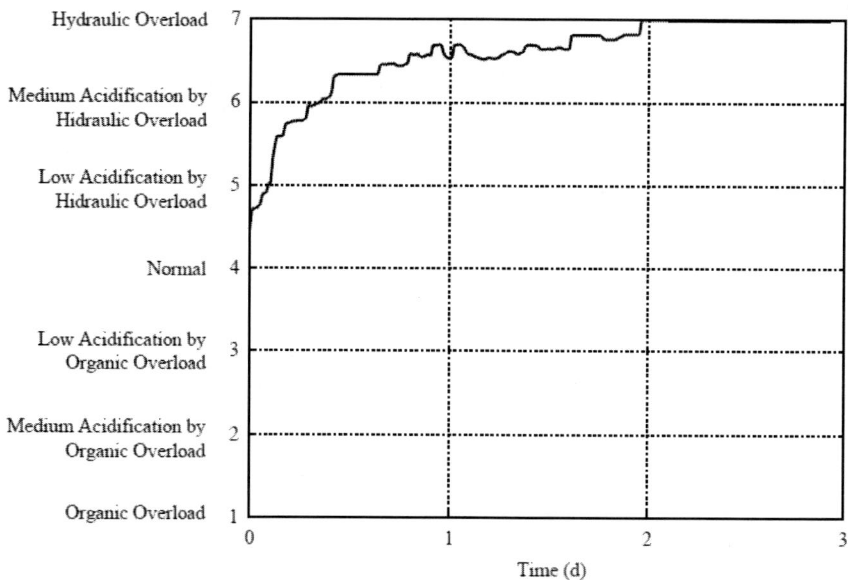

Figure 4. Results obtained from the diagnosis.

The fuzzy diagnosis system was able to let an operator know what an expert would say about the state of the process (Figure 4), in view of the values of five on-line measured variables. From this point of view, the expert system introduces a very useful tool in the overall control of the process, being able to give valuable information to the operators. Implementation of this fuzzy system in a closed-loop control architecture, could also avoid further acidifications of the system due to either hydraulic or organic overloads, by means of acting over pumps or valves being the problem corrected or its consequences minimized.

More recently, the diagnosis system was improved and a closed-loop control system was designed. In this new stage, only three variables (methane production, hydrogen concentration in the gas phase, and intermediate alkalinity/total alkalinity ratio) were selected as the more adequate for organic overload identification, according to previous works (Castellano *et al.*, 2007; Molina *et al.*, 2009). The improved diagnosis and control system was developed using expert's knowledge related with normal or overload process states for establishing a set of rules considering the selected variables. Diluted wine and a dextrin solution were used as models of winery wastewater and rich carbohydrates wastewater, respectively. The wastewater containing carbohydrates consisted of dextrin (chain of 6-8 glucose molecules) dissolved in tap water.

The validation of this improved diagnosis and control system was carried out by simulation with ADM1, open-loop control with experimental data from an anaerobic USBF (hybrid upflow sludge bed-filter) pilot plant, and finally, by closed- loop control. ADM1 is a complex model of the multistep anaerobic process transformations. This tool is adequate for predictions of enough accuracy to be useful in process development, optimization, and control. It is a standard benchmark for developing operational strategies and evaluating controllers which considers processes such as hydrolysis of particulates, acidogenesis, acetogenesis, and methanogenesis, and includes 26 dynamic state concentration variables, 19 biochemical kinetic processes, 3 gas-liquid transfer kinetic processes, and 8 implicit algebraic variables per liquid vessel.

The input variables were IA/TA ratio, methane production (QCH_4), hydrogen concentration in gas phase (H_2), and hydrogen concentration change velocity (DH_2), which may indicate a positive trend (increase) or negative trend (decrease). The control action is established over the feed flow rate (FF), thus the output variable for controlling the system is the variation (increase or decrease) of the feed flow rate (DeltaFlowrate).

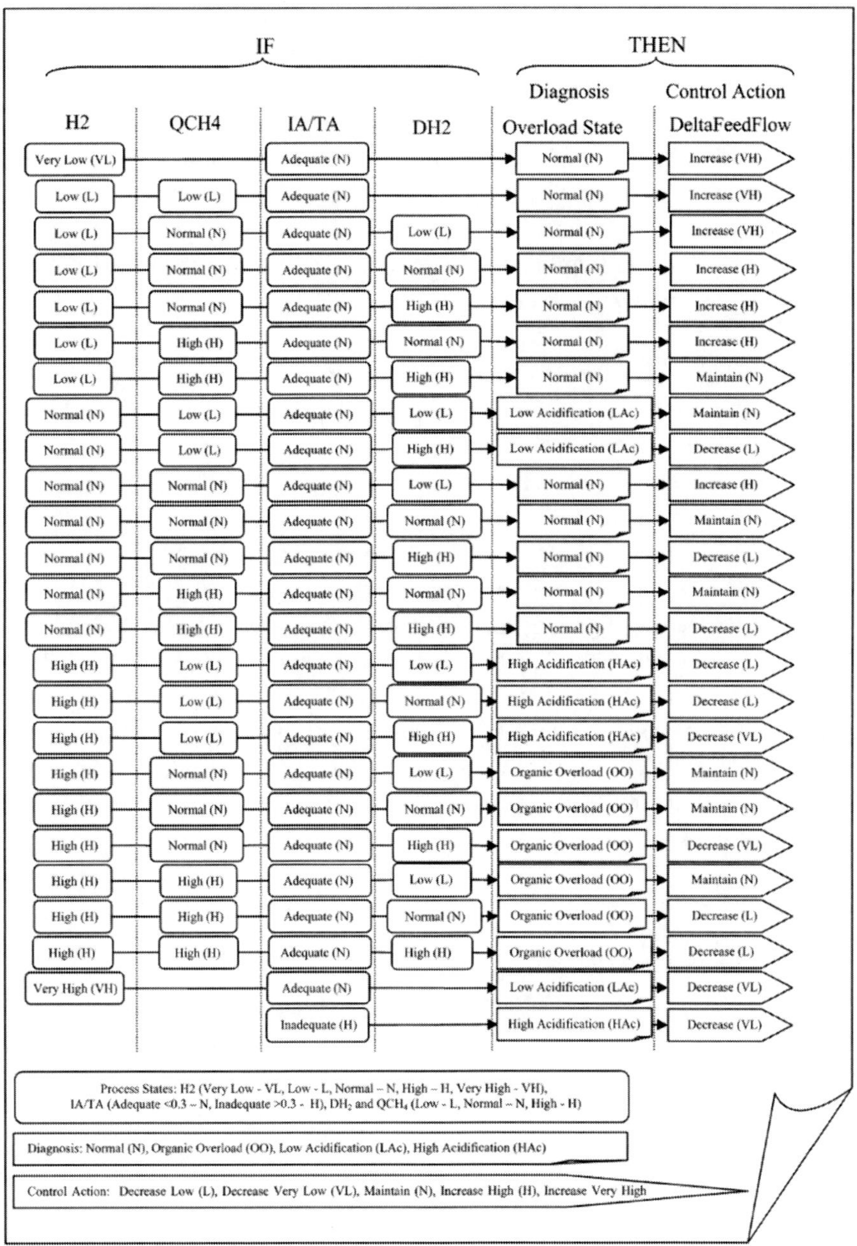

Figure 5. Decision trees for building the knowledge-based rules for diagnosis and control.

The development of the diagnosis and control system was carried out in a first approach, building a knowledge base. This knowledge base was built by compiling the necessary knowledge from the operation of the USBF reactor. It consists of 14 rules for diagnosis and 25 for control, plus 2 constraints in order to prevent hydraulic overloads (criterion of minimum hydraulic retention time of 12 h) and non-negative values for feed flow rate (Figure 5).

The operation states considered for diagnosis were arranged according to the degree of process destabilization by organic overload (from normal to high acidification) as follows:

- Normal (N): without symptoms of organic overload.
- Organic overload (OO): with early symptoms of beginning organic overload.
- Low acidification (LAc): with clear symptoms of acidification by organic overload.
- High acidification (HAc): high degree of acidification.

As can be seen in Figure 5, the set of rules (knowledge base) determines, based on the acquired data, the state of the process and adjusts the feed flow rate. In the particular case shown in Figure 6, the system diagnoses the beginning of an organic overload and establishes a proper control action (a low decrease of feed flow rate) for managing the process and avoiding the progressive acidification of the reactor.

IF

IA/TA is adequate (N)

and H2 is High

and QCH4 is High

and DH2 is zero (N)

THEN (Diagnosis)

Diagnosis is Organic Overload (OO)

AND (Control)

Delta Feed Flow is decreased Low (L)

Figure 6. Example of knowledge-based rule for diagnosis and control.

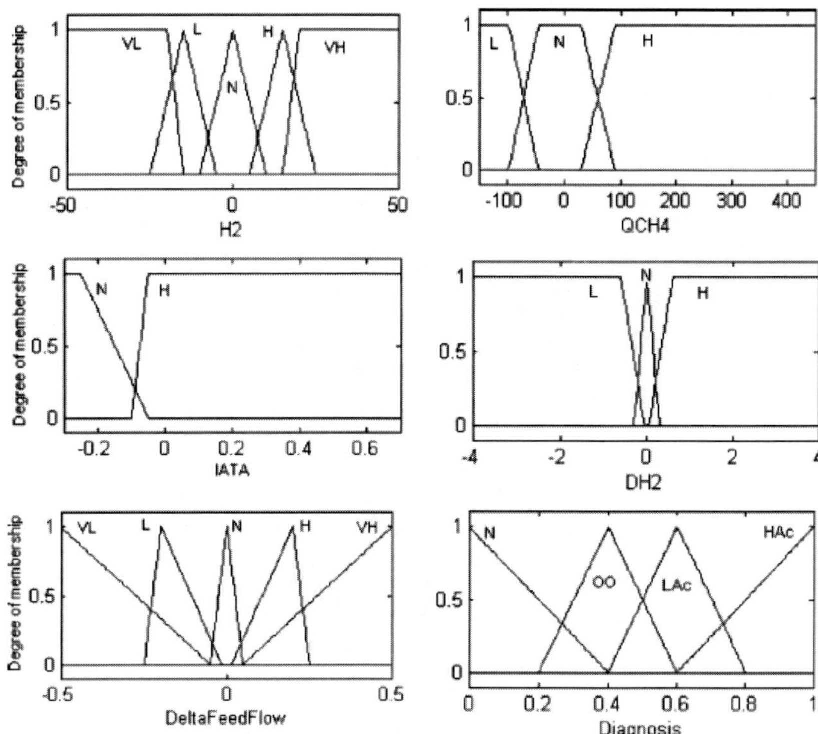

Figure 7. Membership functions for the input variables (H_2, ppm; QCH_4, L/h; IA/TA and DH_2, ppm/min), control actions (DeltaFeedFlow, L/h) and diagnosis of process state: VL, very low; L, low; N, normal; H, high; VH, very high; OO, organic overload; LAc, low acidification; and Hac, high acidification.

The membership functions (Figure 7) were built considering the deviation of the actual value of an input or control variable from a normal value. Therefore, zero is the central value at normal conditions, and negative or positive values are below or above normal conditions, respectively. The membership function for the diagnosis was built according to the process states commented above.

The response surfaces that constitute the defuzzification step for the outputs of the diagnosis and the control action are shown in Figure 8. The zone considered as set point for control action is also indicated. Besides the simple structure and easy implementation, nonlinear variation of the response for H_2 and QCH_4 is one of the advantages of this controller when compared with classical controllers (e.g. PID) based on the same process variables. The system of diagnosis and control also permits the operation in supervisor mode if the controller does not work properly or the reactor is in serious risk. In this mode, the output of the supervisor is the degree of overload applied to the process and the recommendations for handling the process.

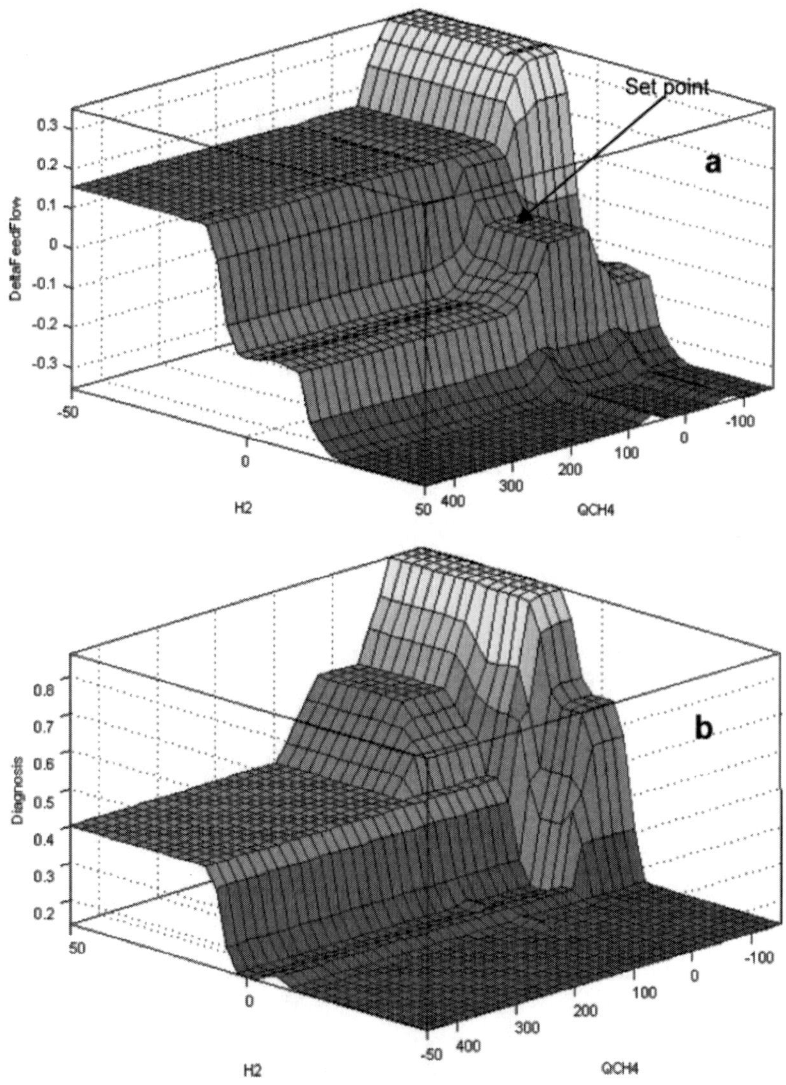

Figure 8. Response surfaces for output defuzzyfication (with normal IA/TA and constant H_2): (a) control, (b) diagnosis; H_2 (ppm), QCH_4 (L/h), and DeltaFeedFlow (L/h).

The developed diagnosis and control system was tested using data generated by simulation with a modified ADM1 commented above; open-loop methodology using experimental data from a reactor treating wastewater containing ethanol; and closed-loop control of a USBF reactor treating wastewater containing carbohydrates.

The validation with modified ADM1 allows the reliability of the system to be tested by simulating two usual industrial scenarios, restart-up and organic overload of the process. The proposed system diagnosed correctly the different episodes, and it demonstrated a high reliability, supplying an adequate control action to achieve the desired set point as well as to manage a sudden change in influent COD. The validation against open-loop data from a USBF reactor treating wastewater containing ethanol allowed the performance of the developed system to be tested considering real data (e.g., H_2 concentration in the gas phase). The control system was able to restart-up the bioreactor and handled the organic overload. Finally, the validation was carried out in a closed-loop control experiment with a different type of wastewater (containing carbohydrates). The controller needed 2 days for restarting-up the process and was able to maintain the process stability during the whole experiment, showing again a high performance and reliability.

2.2. Case Study II: Ecodesign Tool

In this second case study, an ecodesign tool integrating the criteria provided by three environmental evaluation methodologies, namely Ecological Footprint (EF), Life Cycle Assessment (LCA) and Environmental Risk Assessment (ERA), is presented. This tool was constructed on the basis of FL reasoning and features.

This idea enabled the decision making at process and product level taking into account the values of the different indicators at a time. The relative importance of each of them was established through the definition of membership functions as inputs to the fuzzy inference reasoning procedure in the case of a specific product. As a result, a Fuzzy EcoDesign Index (FEcoDI) was obtained. A well-known example was used to test the tool. Different packaging materials for a beverage bottle were considered to identify the most environmentally friendly option (Herva et al., 2012).

2.2.1. Underlying Concepts

Ecodesign may be defined as the systematic introduction of environmental concerns during product design and development (AENOR, 2003). This means to bear in mind the environmental impacts at all stages of the product life cycle, starting at the designing and development phases. The identification and appraisal of the environmental burdens requires the application of evaluation tools. The different environmental available indicators offer complementary visions of the studied scenario; therefore, they cannot be replaced by each other and, in most cases, more than one should be applied at a time (Herva et al., 2011).

As mentioned before, the developed ecodesign tool was constructed on the basis of three environmental evaluation methodologies: EF, LCA and ERA, which are meant to offer complementary perspectives for the evaluation of production processes and products. The EF provides valuable information about the degree of sustainability of a particular process since this indicator especially accounts for resources and energy consumption. However, some limitations were acknowledged for this methodology even though active development on EF methodology poses to continuous new proposals to overcome core critiques (Venetoulis and Talberth, 2008; Kitzes et al., 2009; Herva et al., 2010). Thus, the EF does not capture most of the impact categories usually applied in life cycle analysis or it does not comprehensively take into account waste and emission flows. As a consequence, it results interesting to complement EF studies with certain LCA indicators (derived from the impact assessment) depending on the case study.

LCA is claimed to offer an integrative assessment of a process; however, the information provided regarding human and ecosystem toxicity, for example, is more incomplete than desirable. This means that it has a limited capacity to predict toxicity effects given that the fate of pollutants is usually not considered, so that the calculated impacts are potential rather than actual (Azapagic and Perdan, 2000). The ERA, on the other hand, provides an established methodology based on the appraisal of different scenarios and events, distribution and transfer routes, exposure pathways, duration and frequency of the events that allows for a more rigorous and exhaustive evaluation. Nevertheless, assessments may need to integrate the risks from the entire life cycle of the chemical or product. Therefore, LCA and ERA are complementary tools that can be integrated.

The application of FL reasoning in this case allowed addressing uncertainty matters and the integration of the different indicators, thus supporting the decission making process. The objective was to obtain a final

Fuzzy EcoDesign Index (FEcoDI) for ranking options from an environmental point of view.

2.2.2. Development of the Tool for the Ecodesign of Products

The case study used to construct the tool and to test it was based on a 2 L bottle of drinking water as functional unit. Two kinds of plastic materials were assessed: PET and PVC. Based on the life cycle inventory of these products (Table 1), EF and complementary LCA were estimated. For ERA, particular acknowledged risk problems were taken into account, like the migration of bisphenol A (Le *et al.*, 2008) and aldehydes in PET (Dabrowska *et al.*, 2003), or vinyl chloride monomer in PVC (Fayad *et al.*, 1997).

Table 1. Inventory data for the PVC and PET systems (Feijoo and Roca, 2005)

			PVC		PET	
Input variables	Raw Materials	Iron ore	0.0118	g	13.75	g
		Limestone	4.8	g	6.75	g
		Sand	0.032	g	0.5	g
		Water	640	g	438	g
		Bauxite	7.11	mg	7.75	Mg
	Energy		1.978	MJ	1.868	MJ
Output variables	Air emissions	CO_2	57.6	g	53	g
		CH_4	0.182	g	0.0925	g
		N_2O	0.0002	g	0.0001	g
		NO_x	0.511	g	0.475	g
		SO_x	0.416	g	0.55	g
		Halon 1301	0.0012	mg	0.0018	Mg
		Metals	0.0438	mg	0.0174	Mg
		Aromatic compounds	1.487	mg	0.087	Mg
		Others	0.0081	g	0.0028	g
	Water emissions	COD	0.0352	g	0.0780	g
		Phosphates	0.0022	g	0.0022	g
		Nitrates	0.0003	g	0.0003	g
		Ammonium	0.0005	g	0.0008	g
		Others	0.0083	g	0.0027	g
	Solid waste		4.16	g	1.03	g

Individual EFs were calculated for each material and energy flow in the inventory data, and then they were aggregated to estimate the total EF of each bottle. This indicator was employed to evaluate the energy and materials consumption in the production process of the product.

Emissions released during the manufacture were evaluated via LCA. In this case, the methodology established in the ISO 14040 standards was applied. During the life cycle impact assessment, only the compulsory characterization phase was considered in the development of the ecodesign framework, using factors from the Centre of Environmental Sciences of the Leiden University (CML) were applied using the following equations:

$$C_j = \sum_i C_{ij} = \sum_i A_i \cdot W_{ij} \tag{1}$$

$$C_{jN} = \frac{C_j}{N_j} \tag{2}$$

where A_i is the amount of emission i released, W_{ij} is the characterization factor for the emission i within the category j, C_{ij} is the contribution of the emission i to the category j, C_j is the characterized value of the category j, N_j is the normalization factor of the category j and $C_{j,N}$ is the category j normalized. C_j units depend on the category considered, whereas $C_{j,N}$ is dimensionless.

Global Warming Potential (GWP) and Acidification Potential (AP) impact categories were incorporated into the EF estimate using absorption factors. For the conversion of CO_2 equivalent emissions into area units, a factor of 1.42 tC ha^{-1} yr^{-1} (Wackernagel and Rees, 1996) was employed; meanwhile, for the transformation of acidifying emissions a critical load of 20×10^{-3} eqH^+ m^{-2} yr^{-1} (Holmberg et al., 1999), which is a general threshold for Europe, was applied.

For the environmental risk estimates, migration rates from the bottle material to water, as well as final concentrations in water for the compounds considered (bisphenol A, vinyl chloride monomer and aldehydes), were found in the literature (Table 2).

The concentration of these compounds in water stored depended mainly on initial concentrations in the bottle material, as well as on temperature and time of storage. Estimations were made under the worst case scenario conditions (i.e., major concentrations reported) and only considering the oral pathway. Thus, during the exposure evaluation phase, the following equation was used to estimate the daily dose due to ingestion of water:

Table 2. Data used for the risk characterization

Compound	Material	Migration rate / concentration in water			RfDs(a) mg kg^{-1} day^{-1}	SF(a) kg day^{-1} mg^{-1}
		Value	Units	Source		
Bisphenol A	PET	0.19±0.13	µg l^{-1}	Le et al., 2008	$5.00 \cdot 10^{-2}$	-
Vinyl monomer	PVC	0.6	µg l^{-1}	Fayad et al., 1997	$3.00 \cdot 10^{-3}$	1.50
Acetaldehyde	PET	60.0±6.0	µg l^{-1}	Dabroska et al., 2003	-	-
Formaldehyde	PET	78.1±7.8	µg l^{-1}	Dabroska et al., 2003	$2.00 \cdot 10^{-1}$	-

(a) Source: (ORNL, 2010).

$$Dose = WIF \cdot CW \quad (3)$$

where *Dose* is expressed in mg kg^{-1} day^{-1}, *WIF* is the human water intake factor (a value of $2.5 \cdot 10^{-2}$ L kg^{-1} day^{-1} was considered) and *CW* is the final concentration in water of each compound expressed in mg L^{-1}.

For the risk characterization, Reference Doses (*RfDs*) for non-carcinogenic effects, and Slope Factors (*SF*) for carcinogenic effects, were used in order to calculate the Hazard Quotient (*HQ*) and the Cancer Risk factor (CR) as stated in the following equations:

$$HQ = \frac{Dose}{RfDs} \quad (4)$$

$$CR = Dose \cdot SF \quad (5)$$

HQ and CR values calculated for the different compounds are summed up for each material to obtain the final indexes. The result must keep under maximum acceptable levels of 1 and $1 \cdot 10^{-4}$, respectively.

The linguistic variables in this case were those corresponding to the indicators derived from the application of the environmental evaluation methodologies explained previously. Given that the most important LCA environmental impacts were GWP and AP (obtained in the normalization phase), and they could be integrated into the EF value, membership functions were defined for CR, HQ and EF. The EF contribution to the ecodesign indicator was measured in terms of EF variation (ΔEF) in relation to a base case; that is to say, the higher the decrease in the EF value, the better.

Triangular and trapezoidal functions were selected in all cases. These are the simplest membership functions and they are formed using straight lines. The triangular function consist of a collection of three points forming a triangle, while the trapezoidal membership function has a flat top and it just really consist of a truncated triangle curve. These straight line membership functions have the advantage of simplicity, but provide detail enough to describe the input variables considered in this case.

The Mamdani inference system (Mamdani and Assilian, 1975) was selected and the Fuzzy Logic Toolbox of Matlab® v.7.7 was used to develop the ecodesign tool. Mamdani's fuzzy inference method is the most commonly seen fuzzy methodology. Besides, Mamdani-type inference expects the output membership functions to be fuzzy sets. On the other hand, Sugeno-type systems can be used to model any inference system in which the output membership functions are either linear or constant.

A total of 23 if-then rules were defined and implemented based on the decision tree presented in Figure 9. Some of these rules are shown in Table 3 as an example.

As output to the inference engine, a Fuzzy EcoDesign Index (FEcoDI) was obtained, which collected the different criteria given by the environmental evaluation methodologies used. The best option, from an overall environmental point of view, will be the one with the highest FecoDI (Figure 10).

Table 3. Typical examples of the if-then rules defined (taken from Herva et al., 2012)

IF	THEN
CR *is* Unacceptable	FEcoDI *is* Unacceptable
CR *is* High *and* HQ *is* High	FEcoDI *is* Unacceptable
CR *is* Low *and* HQ *is* High *and* ΔEF *is* Average *or* Bad	FEcoDI *is* Bad
CR *is* Low *and* HQ *is* Low *and* ΔEF *is* Bad	FEcoDI *is* Average
CR *is* Low *and* HQ *is* Medium *and* ΔEF *is* Good	FEcoDI *is* Good

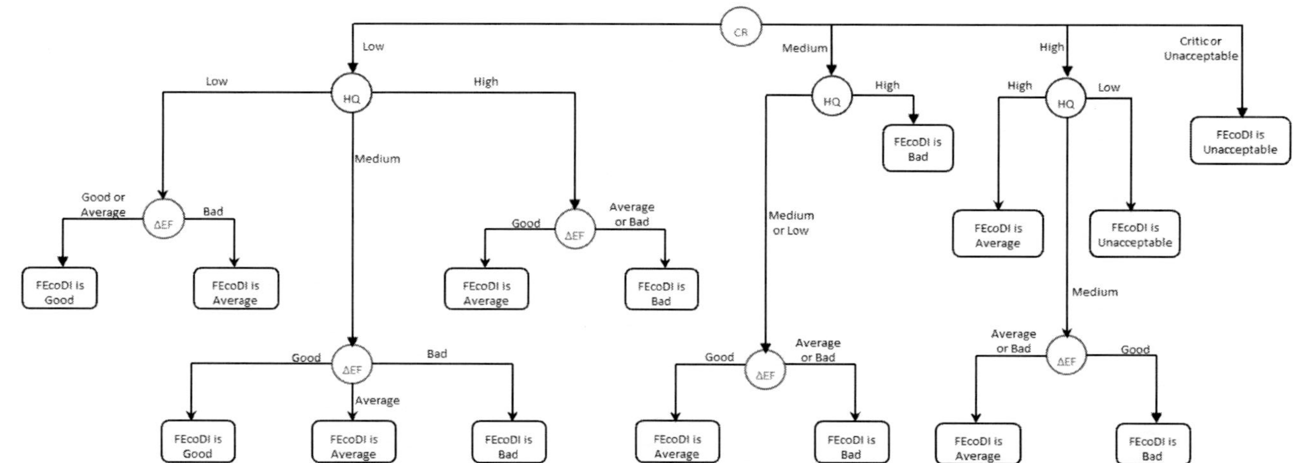

Figure 9. Decision tree defined for the ecodesign tool.

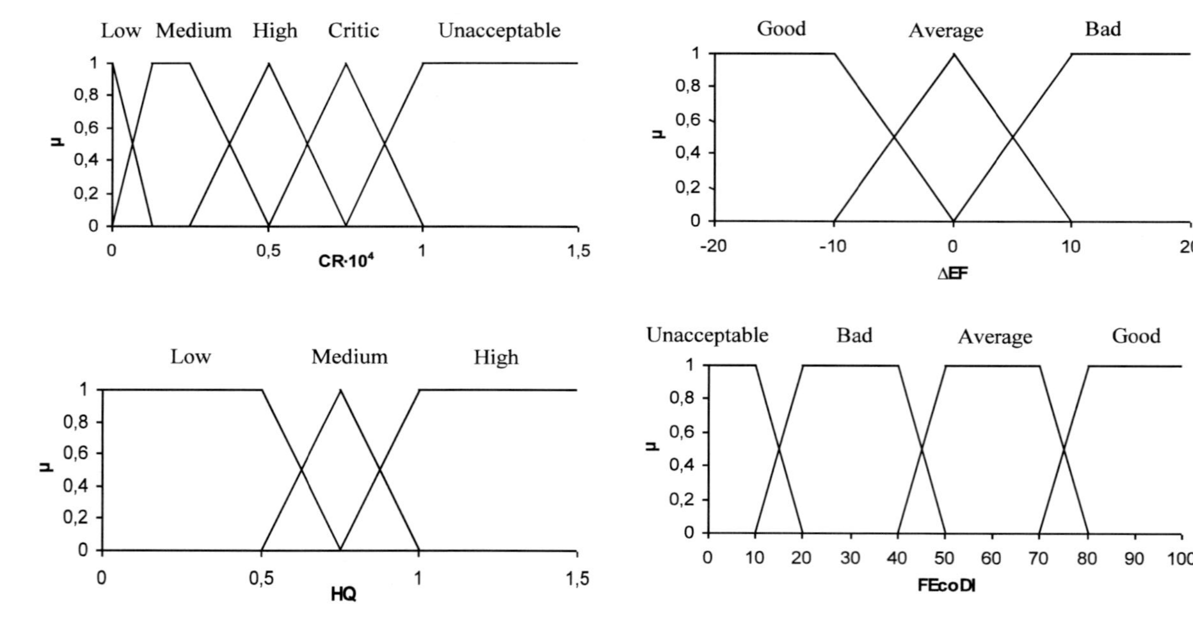

Figure 10. Membership functions for input variables (CR, HQ, EF) and for the output FEcoDI (taken from Herva et al. 2012).

On the basis of the inventoried data for the manufacture of a two liter bottle of these materials, the EF and ACV were calculated, the former accounting for materials and energy consumption, and the latter for emissions to air and water. LCA results for the characterization and normalization phase are shown in Table 4 and Figure 11, respectively. Observing the major importance of GWP and AP categories, and the low relevance of Ozone Depletion Potential (ODP) and Eutrophication Potential (EP) categories, the latter were discarded from this first approach of the ecodesign tool. After the integration of these two into the EF estimates, the total amount of productive land required resulted in 2.27 m^2/bottle for PVC and 2.43 m^2/bottle for PET. Since in the fuzzy structure EF is measured in terms of variance with respect to a base case, the lowest value of the two (PVC) was considered as reference level. Consequently, the input for the PET bottle would correspond to a 7% increase in relation to the EF of the PVC bottle.

Results from ERA were collected in Table 5. Although acetaldehyde and formaldehyde have the R40 risk phrase associated (ESIS, 2010), this meaning that there is limited evidence of a carcinogenic effect (they are catalogued as carcinogenic type 3), currently there is no slope factor available recognized by and international organization for these substances. Furthermore, acetaldehyde has neither *RfDs* nor *SF*; thus, it was not possible to assess the potential environmental risk contribution of this compound. The HQ and CR were below the safety limits stated by the US-EPA in all cases. It must be noticed that in the case of PET the risk contribution of two compounds was considered, while for the PVC only the potential risk derived from the vinyl monomer migration was appraised. This was made according to main risk problems acknowledged in the literature, and taking into account that the purpose of the analysis was to test the tool, not to present definite results on the evaluation of the two materials.

Table 4. Characterization phase for PVC and PET bottle (taken from Herva et al. 2012)

Impact category	Units	PVC	PET
GWP	g CO_2	61.5	55.0
AP	g SO_2	0.75	0.90
ODP	g CFC	$1.44 \cdot 10^{-5}$	$2.16 \cdot 10^{-5}$
EP	g PO_4	$6.96 \cdot 10^{-2}$	$6.60 \cdot 10^{-2}$

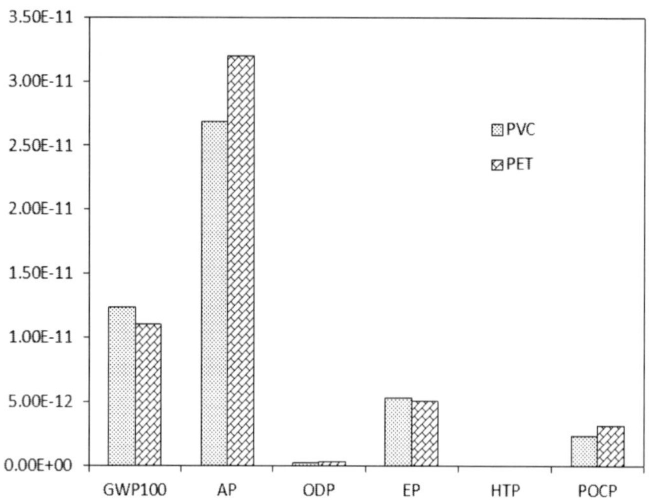

Figure 11. Normalization phase for PVC and PET bottle (taken from Herva et al. 2012).

Table 5. Results for the Environmental Risk Assessment (taken from Herva et al., 2012)

Material	Compound	HQ	CR
PET	Bisphenol A	$9.50 \cdot 10^{-5}$	-
	Formaldehyde	$9.76 \cdot 10^{-3}$	-
	Total	$9.86 \cdot 10^{-3}$	-
PVC	Vinyl monomer	$5.00 \cdot 10^{-3}$	$2.25 \cdot 10^{-5}$

The results obtained applying each environmental evaluation methodology were introduced into the ecodesign tool. As a result, a FEcoDI of 30.0 for the PVC bottle was obtained, while a value of 66.6 was estimated for the PET bottle. Consequently, this would lead to select the PET bottle as the best option from an environmental point of view. The tool was more sensitive to changes in the CR and HQ because priority was given to these variables in the construction of the decision tree (Figure 9). Therefore, materials that may cause carcinogenic effects would receive a bad evaluation from the tool. Hence, in this case, the PET bottle obtained a better FEcoDI in spite of having a higher EF and HQ. Nevertheless, an expansion in the number of linguistic values assigned to these variables could be considered if a major distinction between products with different risk characteristics under the safety limits

(HQ = 1) or more sensitiveness to changes in EF were required. The tool is opened to do the necessary arrangements in the membership functions and decision rules.

CONCLUSION

In this chapter the application of FL techniques to two types of environmental systems was presented.

In the first case study, a FL based system for diagnosis and control of an anaerobic wastewater treatment plant was developed. The controller was built by selecting methane production, hydrogen concentration in the gas phase, and intermediate alkalinity/total alkalinity ratio as control variables. The diagnosis and control system was developed using expert's knowledge related with normal or overload process states for establishing a set of rules considering the selected variables. The controller computes the feed flow rate for adjusting the organic load of the reactor, applying a control law based on a response surface obtained according to fuzzy inference Mamdani methodology.

The validation with modified ADM1 allows the reliability of the system to be tested by simulating two usual industrial scenarios, restart-up and organic overload of the process. The proposed system diagnosed correctly the different episodes, and it demonstrated a high reliability, supplying an adequate control action to achieve the desired set point as well as to manage a sudden change in influent COD. The validation against open-loop data from a USBF reactor treating wastewater containing ethanol allowed the performance of the developed system to be tested considering real data (e.g., H_2 concentration in the gas phase). The control system was able to restart-up the bioreactor and handled the organic overload.

Finally, the validation was also carried out in a closed-loop control experiment with a different type of wastewater (containing carbohydrates). The controller needed 2 days for restarting-up the process and was able to maintain the process stability during the whole experiment, showing again a high performance and reliability.

In the second case study, a first approach of a tool based on EF, LCA and ERA was presented on the basis of FL reasoning and features. The output obtained was a Fuzzy EcoDesign Indicator (FEcoDI) that could range between 0 and 100, and that collected the criteria offered by the different environmental evaluation methods. In general terms, the constructed tool seemed to work properly when tested with the case study.

REFERENCES

AENOR (2003). Environmental Management of Design and Development Process. Design for Environment, UNE 150301 Standards (in Spanish).

Andriantiatsaholiniaina, L. A.; Kouikoglou, V. S.; Phillis, Y. A. (2004). Evaluating strategies for sustainable development: fuzzy logic reasoning and sensitivity analysis. *Ecological Economics, 48*, 149-172.

Azapagic, A., & Perdan, S. (2000). Indicators of sustainable development for industry: A General Framework. *Process Safety and Environmental Protection, 78(4)*, 243-261.

Batstone, D.J.; Keller, J.; Angelidaki, I.; Kalyuzhnyi, S.V.; Pavlostathis, S.G.; Rozzi, A.; Sanders, W.T.M.; Siegrist, H.; Vavilin; V.A. The IWA Anaerobic Digestion Model No 1 (ADM1). *Water Science and Technology, 45(10)*, 65–73

Bellman, R.E.; Zadeh, L. A. (1970). Decision-making in a fuzzy environment. *Management Science, 17*, 141-163.

Bernard, O.; Hadj-Sadok, Z.; Dochain, D.; Genovesi, A.; Steyer, J. P. (2001). Dynamical model development and parameter identification for anaerobic wastewater treatment processes. *Biotechnology and Bioengineering, 75(4)*, 424–439.

Carrasco, E. F.; Rodríguez, J.; Puñal, A.; Roca, E.; Lema, J. M. (2002). Rule-based diagnosis and supervisión of a pilot-scale wastewater treatment plant using fuzzy logic techniques. *Expert Systems with Applications, 22*, 11-20.

Carrasco, E. F.; Rodríguez, J.; Puñal, A.; Roca, E.; Lema, J. M. (2004). Diagnosis and acidification states in an anaerobic wastewater treatment plant using a fuzzy-based expert system. *Control Engineering Practice, 12*, 59-64.

Castellano, M.; Ruiz-Filippi, G.; González, W.; Roca, E.; Lema, J.M. (2007). Selection of variables using factorial discriminant analysis for the state identification of an anaerobic UASB–UAF hybrid pilot plant, fed with winery effluents. *Water Science and Technology, 56(2)*, 139-145.

Dabrowska, A.; Borcz, A.; Nawrocki, J. (2003). Aldehyde contamination of mineral water stored in PET bottles. *Food Additives and Contaminants, 20(12)*, 1170–1177.

ESIS (2010). European chemical substances information system. http://ecb.jrc.it/esis/, Last accessed June 2010.

Fayad, N. M.; Sheikheldin, S. Y.; AlMalack, M. H.; ElMubarak, A. H.; Khaja, N. (1997). Migration of vinyl chloride monomer (VCM) and additives into

PVC bottled drinking water. *Journal of Environmental Science and Health, Part A, 32*, 1065-1083.

Feijoo, G.; Roca, E. (2005). Environmental Technologies. In Hercules de Ediciones, A Coruña (Ed.), *Galicia Ecology*, vol. XLVII: Environmental Science and Technology I. Part II: Technology for environmental protection (pp. 403-427), Spain (in Spanish).

García, C.; Molina, F.; Roca, E.; Lema, J. M. (2007). Fuzzy-based control of an anaerobic reactor treating wastewaters containing ethanol and carbohydrates. *Industrial and Engineering Chemistry Research, 46*, 6707-6715.

Giraldo-Gómez, E.; Duque, M. (1998). Automatic start-up of a high rate anaerobic reactor using a fuzzy logic control system. *5th Latinamerican Workshop-Seminar Wastewater Anaerobic Treatment.* Viña del Mar, Chile.

Herva, M.; Hernando, R.; Carrasco, E.F.; Roca, E. (2010). Development of a methodology to assess the footprint of wastes. *Journal of Hazardous Materials, 180*, 264–273.

Herva, M.; Franco-Uría, A.; Carrasco, E.F.; Roca, E. (2011). Review of corporate environmental indicators. *Journal of Cleaner Production, 19*, 1687-1699.

Herva, M.; Franco-Uría, A.; Carrasco, E. F.; Roca, E. (2012). Application of fuzzy logic for the integration of environmental criteria in ecodesign. *Expert Systems with Applications, 39*, 4427-4431.

Holmberg, J., Lundpvist, U., Robèrt, K.-H., Wackernagel, M. (1999). The Ecological footprint from a systems perspective of sustainability. *International Journal of Sustainable Development and World Ecology, 6*, 17-33.

Kuo, T. C.; Wub, H. H.; Shieh, J. I. (2009). Integration of environmental considerations in quality function deployment by using fuzzy logic. *Expert Systems with Applications, 36*, 7148-7156.

Le, H. H.; Carlson, E. M.; Chua, J. P.; Belcher, S. M. (2008). Bisphenol A is released from polycarbonate drinking bottles and mimics the neurotoxic actions of estrogen in developing cerebellar neurons. *Toxicology Letters, 176*, 149–156.

Lettinga, G. (1995). Anaerobic digestion and wastewater treatment systems. *Antonie van Leeuwenhoek, 67(1)*, 3–28.

Mamdani, E. H., & Assilian, S. (1975). An experiment linguistic synthesis with a fuzzy logic controller. *International Journal on Man-Machine Studies, 7(1)*, 1-13.

Marchini, A.; Facchinetti, T.; Mistri, M. (2009). F-IND: A framework to design fuzzy indices of environmental conditions. *Ecological Indicators, 9*, 485–496.

Molina, F.; Castellano, C.; García, C.; Roca, E.; Lema, J.M. (2009). Selection of variables for on-line monitoring, diagnosis, and control of anaerobic digestion processes. *Water Science and Technology, 60(3)*, 615-622.

Müller, A.; Marsilli-Libelli, S.; Aviasidis, A.; Lloyd, T.; Kroner, S.; Wandrey, C. (1997). Fuzzy control of disturbances in a wastewater treatment plant. *Water Research, 31(12)*, 3157-3167.

Ocampo-Duque, W.; Ferré-Huguet, N.; Domingo, J. L.; Schuhmacher, M. (2006). Assessing water quality in rivers with fuzzy inference systems: A case study. *Environment International, 32*, 733–742.

ORNL (2010). Risk Assessment Information System (RAIS). Oak Ridge National Laboratory, Oak Ridge, TN, USA. http://rais.ornl.gov/, Accessed June 2010.

Phillis, Y. A.; Andriantiatsaholiniaina, L. A. (2001). Sustainability: an ill-defined concept and its assessment using fuzzy logic. *Ecological Economics, 37*, 435-456.

Pullammanappallil, P. D.; Svoronos, S. A.; Chynoweth, D. P.; Lyberatos, G. (1998). Expert systems for control of anaerobic digesters. *Biotechnology and Bioengineering, 58(1)*, 13–22.

Puñal, A.; Rodríguez, J.; Franco, A.; Carrasco, E. F.; Roca, E.; Lema, J. M. (2001). Advanced monitoring and control of anaerobic wastewater treatment plants: Diagnosis and supervision by a fuzzy-based expert system. *Water Science and Technology, 43(7)*, 191.

Sadiq, R.; Husain, T. (2005). A fuzzy-based methodology for an aggregative environmental risk assessment: A case study of drilling waste. *Environmental Modelling and Software, 20*, 33–46.

Siegrist, H.; Vogt, D.; Garcia-Heras, J.L.; Gujer, W. (2002). Mathematical Model for Meso- and Thermophilic Anaerobic Sewage Sludge Digestion. *Environmental Science and Technology, 36(5)*, 1113–1123.

Steyer, J. P.; Buffiere, P.; Rolland, D.; Moletta, R. (1999). Advanced control of anaerobic digestion processes through disturbances monitoring. *Water Research, 33(9)*, 2059–2068.

Wackernagel, M., & Rees, W. (1996). *Our Ecological Footprint: Reducing human impact on Earth.* New Society Publishers, Canada, (Spanish edition 2001, Colección ecología & Medio).

Zadeh, L. A. (1965). Fuzzy sets. *Information and Control, 8*, 338-353.

In: Fuzzy Logic
Editor: Dinko Vukadinovic

ISBN: 978-1-62417-151-2

Chapter 3

ENERGY PLANNING STRATEGY BASED ON METAHEURISTIC OPTIMIZATION TECHNIQUES CONSIDERING FACTS TECHNOLOGY

Belkacem Mahdad*
Department of Electrical Engineering,
Biskra University, Algeria

ABSTRACT

Multi-objective energy dispatch considering practical generator constraints coordinated with multi shunt FACTS technology is a challenging task for utilities for maintaining stability, security and reliability of dynamic power system. In this chapter two methaheuristic optimization methods have been proposed and applied to solve many practical problems related to power system operation and control. The first proposed method called gravitational search algorithm adapted to improve the solution of dynamic economic dispatch considering practical generator constraints (valve point effect and ramp rate limits). A new variant based PSO is proposed to solving multi objective optimal power flow considering distributed shunt FACTS technology, a dynamic mechanism search for PSO parameters in coordination with hierarchical decomposed strategy is proposed to enhance the performances of the

* bemahdad@yahoo.fr.

original PSO method in term of solution quality and convergence characteristic.

The performance of the two proposed approaches has been tested with many practical electrical power system, solving the dynamic economic dispatch with 5 and 10 units considering valve point effect, ramp rate limits and real power transmission losses, solving the combined economic emission dispatch on IEEE 30-Bus, solving the security OPF to the standard IEEE 57-Bus considering multi shunt FACTS devices based static var compensators (SVC), and applied to solving large static economic dispatch, 40 generating units considering valve point effect. The results of the proposed algorithms compared with recent global optimization methods. It is observed that the proposed approaches are capable to finding the near global solution of non-linear and non-differentiable objective functions and obtain a competitive solution at a reasonable time.

Keywords: Energy planning, optimal power flow, dynamic economic dispatch, hierarchical decomposed network, gravitational search algorithm, emission, reactive power planning, FACTS, valve point effect

I. Introduction

Practical optimal power flow strategy consists in determining the output of generation of each unit with respect to predicted load demand over a specified period of time, while satisfying the practical unit constraints and power system security at normal and abnormal situation (overloading and contingency).

In the literature a large number of papers based conventional methods [1-2-3] proposed to solve the standard OPF problem, in general the difficulties associated with using traditional mathematical optimization methods on solving complex and large-scale engineering problems have contributed to the development of alternative solutions. Authors in [4] provides a valuable introduction and surveys the conventional optimization methods.To overcome the problems related to conventional methods such as sensitive to the initial starting points, and restriction on the nature of cost curves, researchers have proposed a second optimization category based evolutionary algorithms for searching near-optimum solutions. Genetic algorithms (GA), simulated annealing (SA), tabu search (TS), and evolutionary programming (EP) [5-6-7], which are the forms of probabilistic heuristic algorithm have been successfully used to overcome the non-convexity problems of the constrained optimal power flow. Very recently a significant review of recent non-deterministic and hybrid methods applied for solving the optimal power flow problem is proposed in [8].

Dynamic active power planning (DAPP) or dynamic economic dispatch (DED) is one of the important OPF sub problem, the main goal of dynamic active power planning is to dispatch the total available power generation units economically over a specified period of time, while satisfying practical constraints such as: valve point effects, prohibited zones, ramp rates (ramp down limit, ramp up limit) these practical constraints affect considerably the life of the rotor of the generating unit.

New methods and many hybrid methods based evolutionary algorithm have been proposed and applied with success to enhance the solution of practical active power planning considering valve point effect, and prohibited zones, like particle swarm optimization (PSO) [9], Ant colony optimization (ACO) [10], Differential evolution (DE) [11], A new robust differential evolution algorithm [12], A new fuzzy adaptive hybrid particle swarm optimization algorithm [13], Improved differential evaluation [14], Harmony search algorithm (HS) [15], Biogeography based optimization method (BOP) [16-17], Modified honey bee mating optimization [18], Artificial bee colony (ABC) [19], Decomposed Parallel GA [20], Differential evolution based dynamic decomposed strategy [21], Quantum-inspired particle swarm optimization [22], Deterministically guided PSO [23], A modified hybrid EP–SQP approach [24], A hybrid multi-agent based particle swarm optimization algorithm [25], Enhanced cross-entropy method [26], Quantum genetic Algorithm [27], Artificial immune system [28], Adaptive particle swarm optimization approach [29], An improved PSO [30], Improved chaotic particle swarm optimization Algorithm [31], An adaptive hybrid differential evolution Algorithm [32], Imperialist competitive Algorithm [33], Gravitational Search Algorithm [34-35-36], Adaptive fuzzy controlled PSO based dynamic decomposed network strategy [37], these methods have a better searching ability in finding near global optimal solution compared to the mathematical methods

In this work two methods proposed to solve many problems related to power system planning and control: The first method called gravitational search algorithm adapted to solving dynamic economic dispatch considering valve point effect and ramp rate limits, the second method is a new variant of PSO named adaptive hierarchical decomposed PSO proposed to enhance the optimal power flow solution considering shunt FACTS technology.

The proposed approaches have been examined and applied to the 5 and 10 generating units considering ramp rates limits and valve point effect, to IEEE

30-Bus for solving the combined economic emissions dispatch, and to the IEEE 57-Bus with smooth cost function considering shunt FACTS devices. From simulation results it is observed that the results compared with the other recent techniques demonstrate the potential of the proposed approaches and show clearly theirs effectiveness to solve practical OPF in term of solution quality and convergence characteristics.

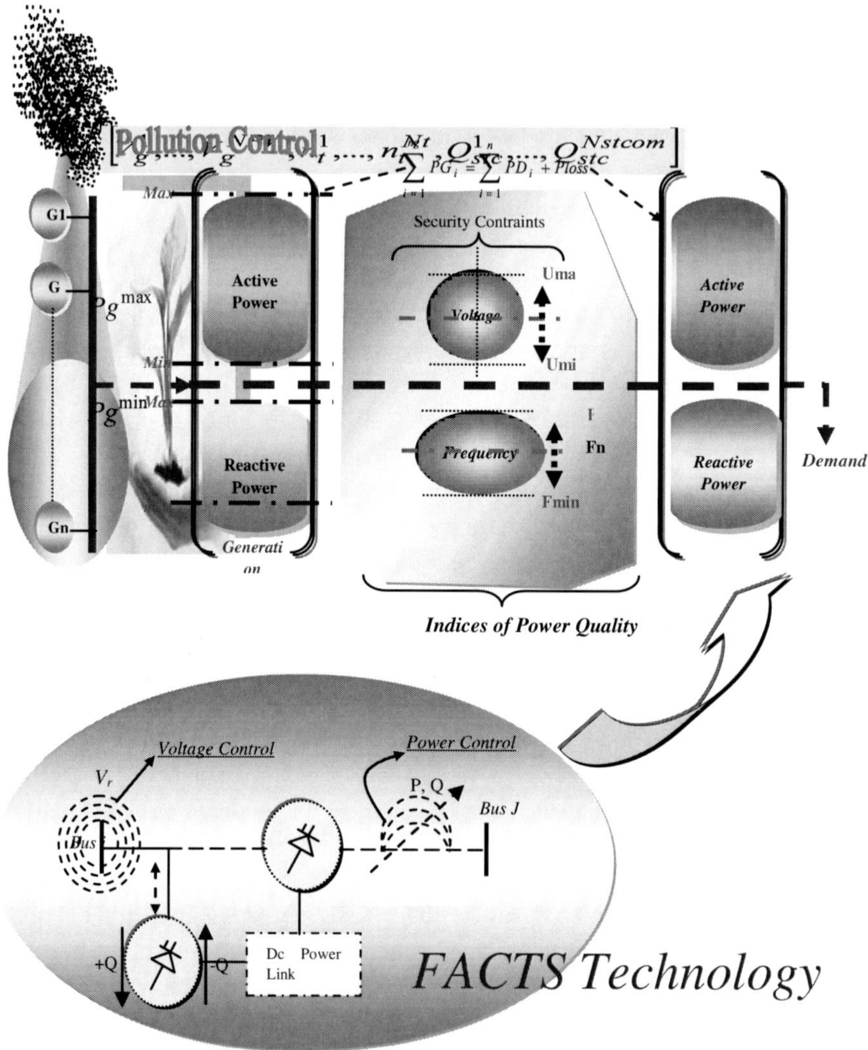

Figure 1. Energy planning strategy considering FACTS technology.

II. Energy Management: Technology and Economic Issue

It is important to underline the importance of energy efficiency management in power systems, the combined term energy strategy is usually associated with how energy is produced (economic aspect), how energy is consumed at the point of end use (technical aspect), and what is the impact of the total energy produced on the environment.

Figure 1 shows the energy planning strategy considering FACTS technology; the following point should be taken by expert engineers and researchers to assure energy efficiency.

1. *Environmental aspect*: The Kyoto protocol can be considered as the first step towards the improvement of energy efficiency by forcing the utilities and expert engineers to introduce the environmental constraints to the standard optimal power flow problem.
2. *Technical aspect:* Consists in considering new technology based FACTS devices and renewable source: The wide integration of renewable energy and flexible ac transmission system (FACTS) in the electric power system becomes very complex and difficult to control efficiently.
3. *Economic aspect*: How to estimate economically the number, the size of FACTS Controllers coordinated with renewable sources to be installed in a practical and large electrical network?

III. Mathematical Formulation of Energy Planning Strategy

The main target of energy planning strategy is to optimize a multi objective function based generation cost, power loss, voltage deviation, voltage stability individually or simultaneously but still satisfying specified constraints (generators constraints and security constraints). The mathematical multi objective problem can be formulated as:

$$\text{Minimize } J_i(x,u) \quad i=1,\dots,N_{obj} \tag{1}$$

$$\text{Subject to: } g(x,u)=0 \tag{2}$$

$$h(x,u)=0 \tag{3}$$

Where J_i is the *ith* objective function, and N_{obj} is the number of objectives. g is the equality constraints, h is the system operation constraints. The vector of state and control variables are denoted by x and u respectively.

1. Dynamic Active Power Planning

1.1. Objective Functions

The objective of dynamic active power planning (DAPP) or dynamic economic dispatch (DED), is to minimize the fuel cost of a specified generating units over the given dispatch time, mathematically, it can be formulated as:

$$\text{Minimize } F_T = \sum_{t=1}^{T}\sum_{i=1}^{NG} F_i(P_{it}) \tag{4}$$

Where F_T is the total fuel cost (\$); NG is the number of generating units; T is the number of intervals in the scheduling horizon; P_{it} is the real power generation of generating unit *i* during sub interval *t* and F_i ()is the fuel cost function of generating unit *i*.

1.1.1. Simple Fuel Cost Function

The simplified objective function F is the total generation cost based quadratic form expressed as follows:

$$\text{Min } F=\sum_{i=1}^{NG}\left(a_i+b_iP_{gi}+c_iP_{gi}^2\right) \tag{5}$$

Where NG is the number of thermal units, P_{gi} is the active power generation of unit *i,* a_i , b_i and c_i are the cost coefficients of the *ith* generator. Figure 2 shows the structure of vector control for active power planning.

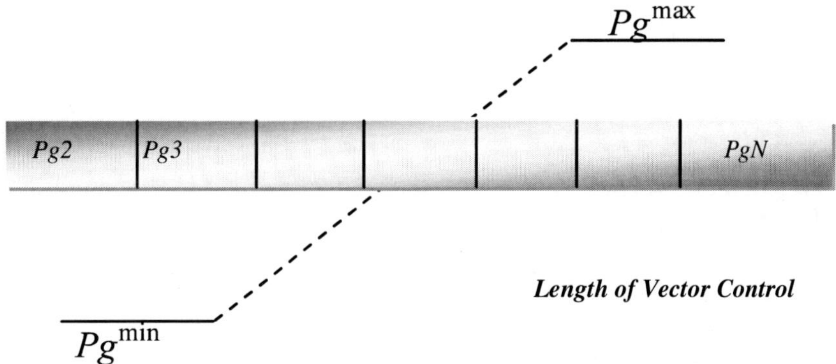

Figure 2. Vector control structure for active power dispatch.

1.1.2. Objective Function Considering Valve Point Loading Effect

The valve-point loading is taken in consideration by adding a sine component to the cost of the generating units [23-24].

Typically, the fuel cost function of the generating units with valve-point loading is represented as follows:

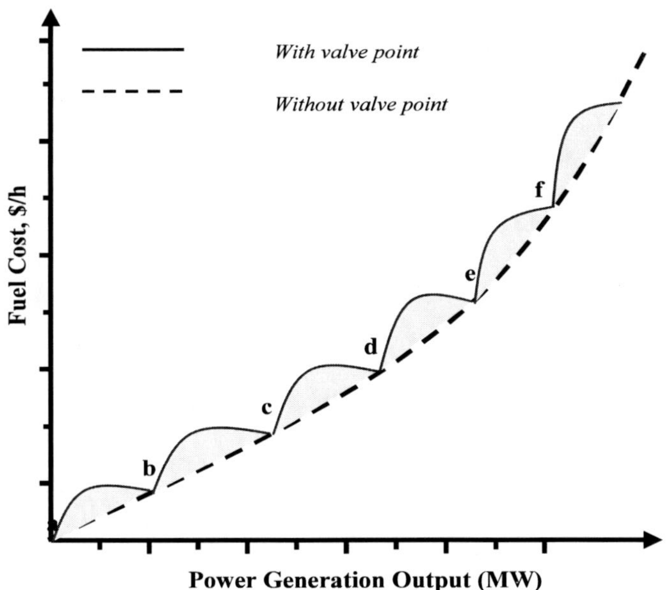

Figure 3. Input-Output curve under valve-point loading effects.

$$f_T = \sum_{i=1}^{NG} \left(a_i + b_i P_{gi} + c_i P_{gi}^2 + \left| d_i \sin\left(e_i \left(P_{gi}^{min} - P_{gi} \right) \right) \right| \right) \tag{6}$$

d_i and e_i are the cost coefficients of the unit with valve point effects.The input-output performance curve for a typical thermal unit can be represented as shown in Figure 3.

1.1.3. Emission Objective Function

The objective function that minimizes the total emissions can be expressed as the sum of all the three pollutants (NO_x , CO_2 , SO_2) resulting from generator real power, this objective function expressed as follows:

$$F_{em} = \sum_{i=1}^{NG} 10^{-2} \times \left(\alpha_i + \beta_i P_{gi} + \gamma_i P_{gi}^2 + \omega_i \exp\left(\mu_i P_{gi} \right) \right) \text{ Ton/h} \tag{7}$$

Where α_i , β_i , γ_i, ω_i and μ_i are the parameters estimated on the basis of unit emissions test results.

2.1. Reactive Power Planning

The main role of reactive power planning (RPP) is to adjust dynamically the control variables individually or simultaneously to reduce the total power loss, transit power flow in branches, voltage deviation, and improve the voltage stability, but still satisfying specified constraints (generators constraints and security constraints). In general three objective functions (power loss, voltage deviation and voltage stability) are considered to be minimized individually or simultaneously. Figure 4 shows the structure of the control variable to be optimized according to the following two objective functions.

2.1.1. Power Loss

The objective function here is to minimize the active power loss (P_{loss}) in the transmission system:

$$J_1 = \text{Min} \left(P_{loss} \right) \tag{8}$$

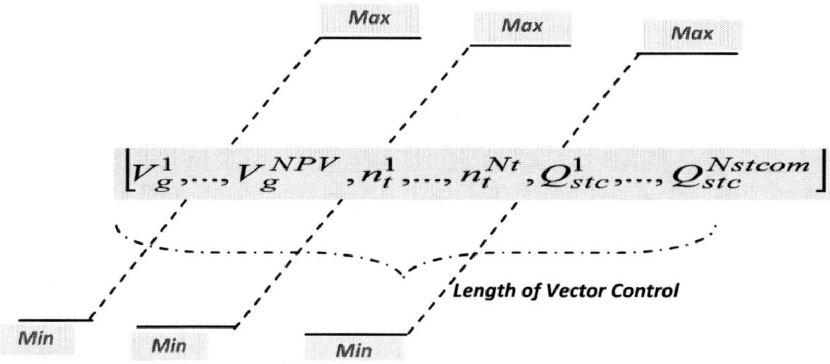

Figure 4. Vector control structure for reactive power planning.

$$P_{loss} = \sum_{k=1}^{N_l} g_k \left[(t_k V_i)^2 + V_j^2 - 2t_k V_i V_j \cos\delta_{ij}\right] \tag{9}$$

Where, V_i, V_j are the voltage magnitudes at bus i, j, respectively; δ_{ij} is the phase angle difference between buses *i* and *j* respectively, g_{ij} and b_{ij} are the real and imaginary part of the admittance (Y_{ij}).

2.1.2. Voltage Deviation

Voltage magnitude is one of the important indices of power quality. For a secure operation of the power system, it is important to maintain required level of security margin. In this case the objective function to be optimized is given as follows:

$$J_2 = \min \sum_{i\in NL} |V_i - 1.0| \tag{10}$$

3. CONSTRAINTS

3.1. Equality Constraints

The equality constraints $g(x)$ are the real and reactive power balance equations, expressed as follows:

$$P_{gi} - P_{di} = V_i \sum_{j=1}^{N} V_j \left(g_{ij} \cos \delta_{ij} + b_{ij} \sin \delta_{ij} \right) \quad (11)$$

$$Q_{gi} - Q_{di} = V_i \sum_{j=1}^{N} V_j \left(g_{ij} \sin \delta_{ij} - b_{ij} \cos \delta_{ij} \right) \quad (12)$$

Where N is the number of buses, P_{gi}, Q_{gi} are the active and the reactive power generation at bus i; P_{di}, Q_{di} are the real and the reactive power demand at bus i; V_i, V_j, the voltage magnitude at bus i, j, respectively; δ_{ij} is the phase angle difference between buses i and j respectively, g_{ij} and b_{ij} are the real and imaginary part of the admittance (Y_{ij}).

3.2. Inequality Constraints

3.2.1. Generator Constraints

The Ramp Rate Limits

To avoid severe perturbation on the boiler and the combustion equipment which affect directly the turbine stability, the real power generation constraint should be superimposed by the ramp-rate constraint; this is mathematically represented as follows:

$$P_{it}^{min} \leq P_i(t) \leq P_{it}^{max} \quad (13)$$

$$P_{it}^{min} = \max\left(P_i^{min}, P_i(t-1) - DR_i\right) \quad (14)$$

$$P_{it}^{max} = \min\left(P_i^{max}, P_i(t-1) + UR_i\right) \quad (15)$$

Where P_{it}^{min}, P_{it}^{max} are the modified minimum and maximum limit of the real power of the *ith* unit at the *tth* interval in MW and $P_i(t-1)$ is the power output of *ith* unit at time (t-1) in MW, UR_i and DR_i are the up and down ramp rate limits of ith unit in MW/h.

3.2.2. Security Limits

The inequality constraints on security limits are given by:

- Constraints on transmission lines loading

  $S_{li} \leq S_{li}^{max}, \ i = 1,2,\dots,NPQ$ (16)

- Constraints on voltage at loading buses (PQ buses)

  $V_{Li}^{min} \leq V_{Li} \leq V_{Li}^{max}, \ i = 1,2,\dots,NPQ$ (17)

- Generator constraints

  Upper and lower limits on the generator bus voltage magnitude:

  $V_{gi}^{min} \leq V_{gi} \leq V_{gi}^{max}, \ i = 1,2,\dots,NPV$ (18)

- Transformer constraints

  Upper and lower limits on the tap ratio (t) of transformer.

  $t_i^{min} \leq t_i \leq t_i^{max}, i = 1,2,\dots,NT$ (19)

- FACTS Controllers constraints

  Parameters of shunt FACTS Controllers must be restricted within their upper and lower limits.

  $X^{min} \leq X_{FACTS} \leq X^{max}$ (20)

IV. GRAVITATIONAL SEARCH ALGORITHM

GSA has been inspired from the mass interactions based on the general gravitational law. In the proposed algorithm, the search agents are a collection of masses, interact with each other based on the Newtonian laws on gravity and motion. Like many metaheuristic optimization methods and to increase diversity and the probability of finding the near global optimum [34], this new method requires the adjustment of some stochastic parameters.The flowchart of the basic GSA is presented in Figure 5.

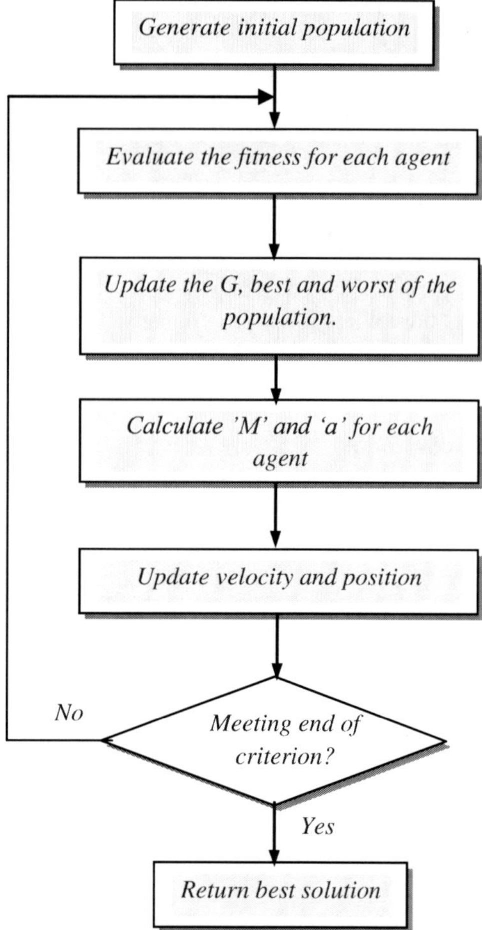

Figure 5. Flowchart of the basic GSA.

1. Brief Description of GSA Mechanism Search

In GSA, each mass (agent) characterized by four specifications: position, inertial mass, active gravitational mass, and passive gravitational mass. The position of the mass corresponds to a solution of the problem, and its gravitational and inertial masses are determined using a fitness function. In the following, the formulation of GSA is presented in short. Consider a system with N agents (masses). We define the position of the *ith* agent by:

$$X_i = \left(x_i^1, \ldots, x_i^d, \ldots, x_i^n\right) \quad \text{for } i = 1, 2, \ldots, N, \tag{21}$$

Where x_i^d presents the position of *ith* agent in the *dth* dimension.

The force applied on mass *i* at iteration *t* is defined by:

$$F_{ij}^d(t) = G(t)\frac{M_{pi}(t).M_{aj}(t)}{R_{ij}(t)}\left(x_j^d(t) - x_i^d(t)\right), \tag{22}$$

Where M_{aj} is the active gravitational mass related to agent *j*, M_{pi} is the passive gravitational mass related to agent *i*, $G(t)$ is the gravitational constant calculated at iteration *t*, and $R_{ij}(t)$ is the Euclidian distance between two agents *i* and *j*.

The total stochastic force that acts on agent *i* in a dimension d can be expressed as follows:

$$F_i^d(t) = \sum_{j=1\, j\neq i}^{N} \text{rand}_j F_{ij}^d(t), \tag{23}$$

Where $rand_j$ is a uniform random number within the interval [0, 1], the acceleration of the agent *i* at iteration *t* is defined as:

$$a_i^d(t) = \frac{F_i^d(t)}{M_{ii}(t)} \tag{24}$$

Where, M_{ii} is the inertial mass of *ith* agent, the velocity and position of the agents are updated as follows:

$$v_i^d(t+1) = \text{rand}_i v_i^d(t) + a_i^d(t) \tag{25}$$

$$x_i^d(t+1) = x_i^d(t) + v_i^d(t+1) \tag{26}$$

At each iteration *t* the mass of each agent *i* updated based on the following equation:

$$m_i(t) = \frac{fit_i(t) - worst(t)}{best(t) - worst(t)}, \quad (27)$$

$$M_i(t) = \frac{m_i(t)}{\sum_{j=1}^{N} m_j(t)} \quad (28)$$

Where $fit_i(t)$ represent the fitness value of the agent i at iteration t, $worst(t)$ and $best(t)$ are the worst and the best fitness defined as follows:

$$best(t) = \min fit_j(t) \quad j \in \{1,...N\} \quad (29)$$

$$worst(t) = \max fit_j(t) \quad j \in \{1,...N\} \quad (30)$$

V. Overview of PSO

Particle swarm optimization (PSO) is a new branch of EA proposed by Eberhart and Kennedy in 1995 [9]. PSO has proven to be promising candidate to solve real valued optimization problem. In the literature many variants based PSO have been proposed to enhance the performance of the original PSO algorithm. The following descriptions summarize the basic mechanism search of PSO:

Let $X_i = (x_{i1}.....x_{in})$, and $V_i = (v_{i1}......v_{in})$ denote the coordinates and the corresponding flight speed of the particle i in a search space, respectively. A particle in a swarm moves to better position with randomly weighted acceleration using its present velocity, previous experience, and the experience exchanged with others particles. The velocity of the particle is changed according to the relative locations of *Pbest* (personal experience) and *Gbest* (experience of others). It is accelerated in the directions of these locations of greatest fitness according to the following equations [30].

$$V_i^{k+1} = \omega V_i^k + c_1 rd1 \times \left(Pbest_i^k - X_i^k\right) + c_2 rd2 \times \left(Gbest^k - X_i^k\right) \quad (31)$$

$$X_i^{k+1} = X_i^k + C.V_i^{k+1} \tag{32}$$

Where

V_i^k : Velocity of particle *i* at iteration *k*;

ω : Inertia weight factor;

c_1, c_2 : Acceleration factors;

X_i^k : position of particle *i* at iteration *k*;

$Pbest_i^k$: Best position of particle *i* until iteration *k*;

$Gbest^k$: Best position of group until iteration *k*;

rd1 , rd2 : two independently generated random numbers between 0 and 1.

C : Constriction factor calculated as follow:

$$C = \frac{2}{2-\phi-\sqrt{\phi^2-4\phi}} \tag{33}$$

Where; $\phi = (c_1 + c_2) \geq 4$

In this standard variant the inertia weight is linearly decreasing as the iterations proceed and calculated based on the following equation:

$$\omega = \omega_{max} - \frac{(\omega_{max} - \omega_{min})}{iter_{max}}.k \tag{34}$$

Where, ω_{max}, ω_{min} are initial and final weights, k is the current iteration, and $iter_{max}$ is the maximum iteration number.

VI. Adaptive Hierarchical Decomposed PSO

1. Hierarchical Decomposed Strategy

Parallel execution of various standard metaheuristic optimization methods (GA, PSO, DE) is called (Parallel Algorithm). Parallelism have been developed to reduce the large execution times that are associated with the standard metaheuristic optimization methods to solving large problems and to find better solutions. Parallelism can easily be implemented on networks of heterogeneous computers or on parallel mainframes.

The way in which original metaheuristic optimization method can be parallelized depends upon the following elements:

- How the numbers of sub systems are chosen?
- How fitness is evaluated?
- How operators related to a specified method are chosen and applied?
- How information is exchanged between sub systems to enhance the solution quality?

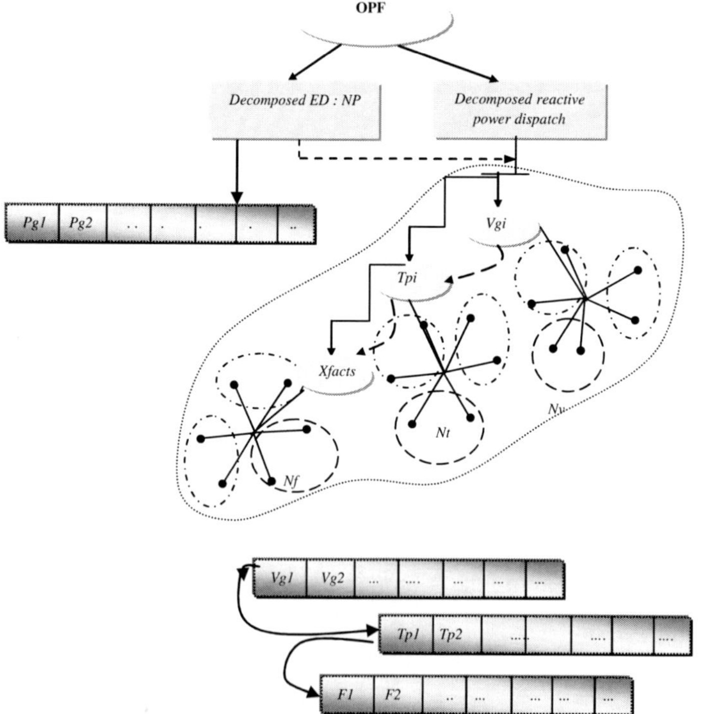

Figure 6. The proposed hierarchical adaptive PSO mechanism search: decomposed active power dispatch & decomposed reactive power dispatch.

In this chapter an improved decomposed approach inspired from the original idea developed in [20] named hierarchical decomposed strategy is proposed to enhance the performance of the original PSO. The multi objective OPF decomposed on two coordinated problems: active power dispatch and reactive power planning. Figure 6 illustrates the flowchart of the proposed hierarchical decomposed strategy applied to solving complex practical OPF. Figure 7 shows the hierarchical adaptive PSO based reactive power dispatch for voltage control.

The following steps summarize the proposed mechanism search:

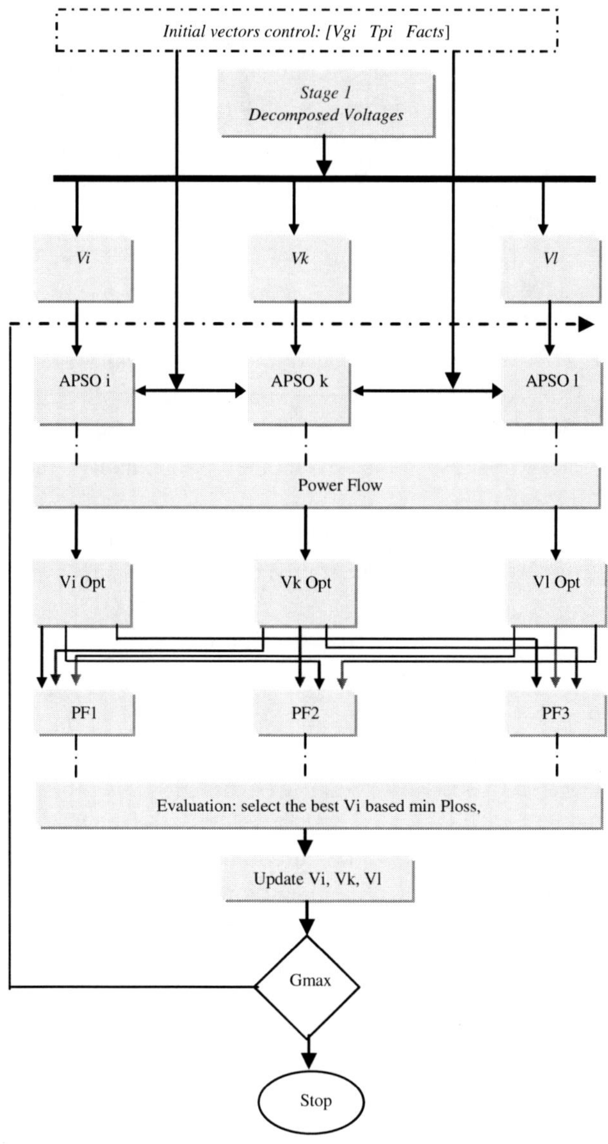

Figure 7. Hierarchical adaptive PSO based reactive power dispatch for voltage control.

1. The vector control based active power generation found in the first stage considered as an initial feasible solution to the second stage.
2. The vector control associated to the reactive power planning composed of voltage generating units, tap transformer setting, reactive power of static compensator, reactive power of shunt FACTS controllers.
   2.1. First: the vector control decomposed on *Ns* subsystems, in our case: Ns=4: voltage control, tap transformer setting, static compensator, and dynamic shunt controllers.
   2.2. Second: voltage control vector is decomposed on *Nv* subsystems.
      - Vector variables based Tap transformer is decomposed on *Nt* subsystems,
      - Vector variables based static compensator is decomposed on *Nc* subsystems,
      - Vector variables based dynamic shunt compensators based SVC controllers is decomposed on *Nf* subsystems.
   2.3. The decomposed subsystems are optimized based on superposition principle, at each level for each decomposed vector; subsystems exchange dynamically their optimized information.
3. Power flow is executed to evaluate the fitness function.
4. Save the best subsystems: total power loss, voltage deviation, total cost.
5. Repeat steps 2.2 to 4 for the tap transformer vector, the static compensator, and for the dynamic shunt controllers:

2. Dynamic Adjustment of PSO Parameters

In this variant based PSO the cognitive and social factors aren't constant but they are adjusted dynamically to explore most all the positions space research to eliminate the local minima. Figure 8 shows the proposed mechanism search based PSO; the PSO parameters are adjusted dynamically during processes search.

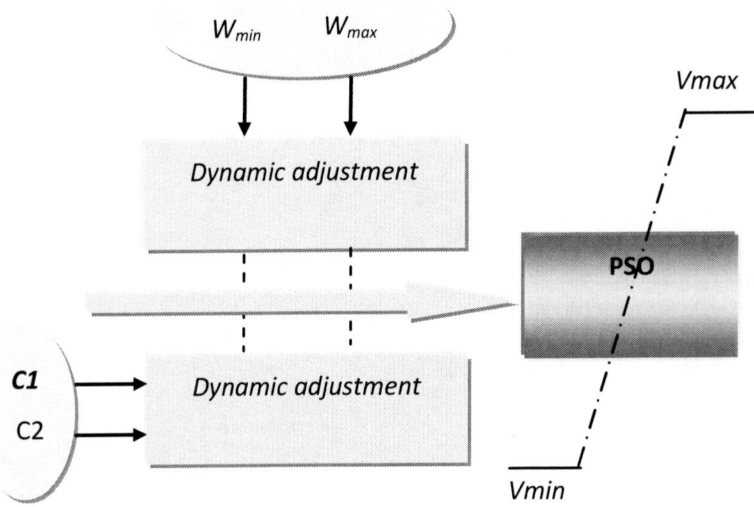

Figure 8. The proposed adaptive PSO mechanism search.

The idea behind the dynamic adjustment of the PSO parameters is to enhance the global search in the early part of the optimization and to encourage the particles to converge towards the global optima at the end of the search.This is achieved by changing dynamically the following parameters:

Cognitive and Social Components

The acceleration coefficients $C1$ and $C2$ are varied dynamically during the search process in such a manner that the cognitive component is reduced while the social component is increased as the search proceeds, Figure 9 shows the dynamic adjustment of acceleration factors for fuel cost and emission.The acceleration coefficients are expressed as:

$$\begin{cases} C1 = \sin\left((C_{1f} - C_{1i})\dfrac{iter}{iter_{max}}\right) + C_{1i} \\ C2 = \sin\left((C_{2f} - C_{2i})\dfrac{iter}{iter_{max}}\right) + C_{2i} \end{cases} \tag{35}$$

Where C_{1f}, C_{1i}, C_{2f} and C_{2i} are initial and final values for social acceleration factors and cognitive respectively.

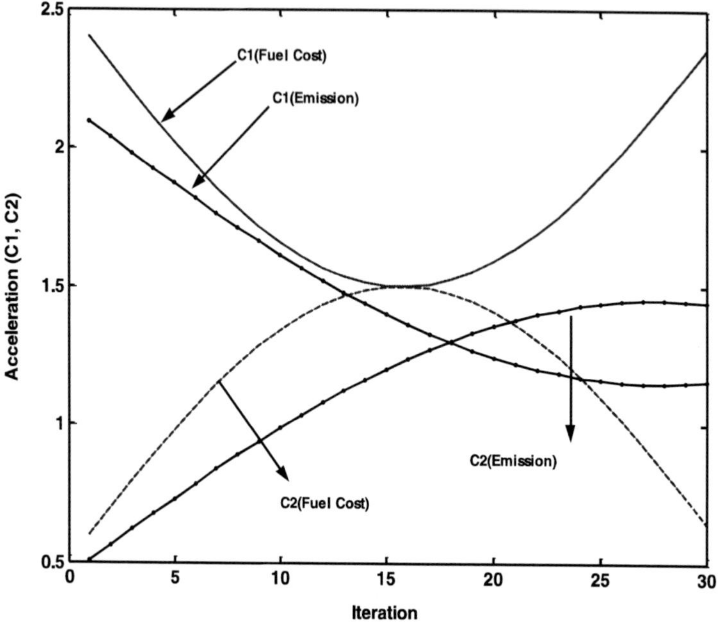

Figure 9. Sample example of dynamic adjustment of acceleration factors for fuel cost and emission.

Particles Movement

A differential factor is added to Eq (20), to improve the performance of the original PSO. The update of particles velocity is formulated as follows.

$$V_i^{k+1} = \omega V_i^k + c_1 rd1 \times \left(Pbest_i^k - X_i^k\right) + c_2 rd2 \times \left(Gbest^k - X_i^k\right) + CT.rd3 \times \left(Gbest^k - Pbest^k\right) \tag{36}$$

$$CT = \begin{cases} 0 & \text{if } iter \le iter_{min} \\ abs(c1 - c2) & \text{if } iter_{min} \prec iter \le iter_{med} \\ abs(c2 - c1) & \text{if } iter_{med} \prec iter \le iter_{max} \end{cases} \tag{37}$$

Where; CT represents a differential coefficient.

$iter_{min}$, $iter_{med}$, and $iter_{max}$ are the minimum, medium, and maximum iteration respectively specified based on a practical experience.

VII. Shunt FACTS Steady State Modelling

Static VAR Compensator (SVC)

The Static VAr Compensator (SVC) [21] is a shunt connected VAr compensator. The SVC has the ability to exchange dynamically reactive power (absorb or generate) with the network to control voltage at specified buses of the electric power system. It includes separate equipment for leading and lagging VArs. The SVC model used in this study is based on representing the shunt Controller as variable susceptance or firing angle control, the following equations describes the SVC model.

$$I_{SVC} = jB_{SVC}V \tag{38}$$

$$\left.\begin{aligned} B_{SVC} &= B_C - B_{TCR} = \frac{1}{X_C X_L}\left\{X_L - \frac{X_C}{\pi}\left[2(\pi-\alpha)+\sin(2\alpha)\right]\right\}, \\ X_L &= \omega L,\ X_C = \frac{1}{\omega C}, \end{aligned}\right\} \tag{39}$$

Where, $B_{SVC}, \alpha, X_L, X_C$, V are the shunt susceptance, firing angle, inductive reactance, capacitive reactance of the SVC controller, and the bus voltage magnitude to which the SVC is connected, respectively.
The reactive power Q_i^{SVC} exchanged with the bus *i* can be expressed as:

$$Q_i^{SVC} = B_i^{SVC}.V_i^2 \tag{40}$$

VIII. Simulation Results

The proposed algorithm is developed in the Matlab programming language (7.6 version) using Microsoft Windows XP. All the programs were run on 2.6 GHz Pentium IV processor with 500MB of random access memory. The proposed approach has been tested on to some standard benchmark cost functions and to many power system tests; 5 generating units and 10 generating units with ramp rate limits and valve loading effect, IEEE 30-Bus, and IEEE 57-Bus test system.

Test 1: Benchmark functions

In this case, the proposed algorithm is tested with many benchmark functions. All these problems are minimization problems. The details of these functions are listed in Table 1, Figs 10-11 show the convergence characteristic of two test functions 'F1' and 'F2' respectively. Due to the limited chapter length, only two functions are considered for minimization problem, details can be retrieved from [34].

Table 1. Unimodal test functions: Dim=30, Itermax=1000

Test Function	Dim	S
$F_1(X)=\sum_{i=1}^{n} x_i^2$	30	$[-100,100]^n$
$F_2(X)=\sum_{i=1}^{n} x_i+\prod_{i=1}^{n}\lvert x_i\rvert$	30	$[-10,10]^n$

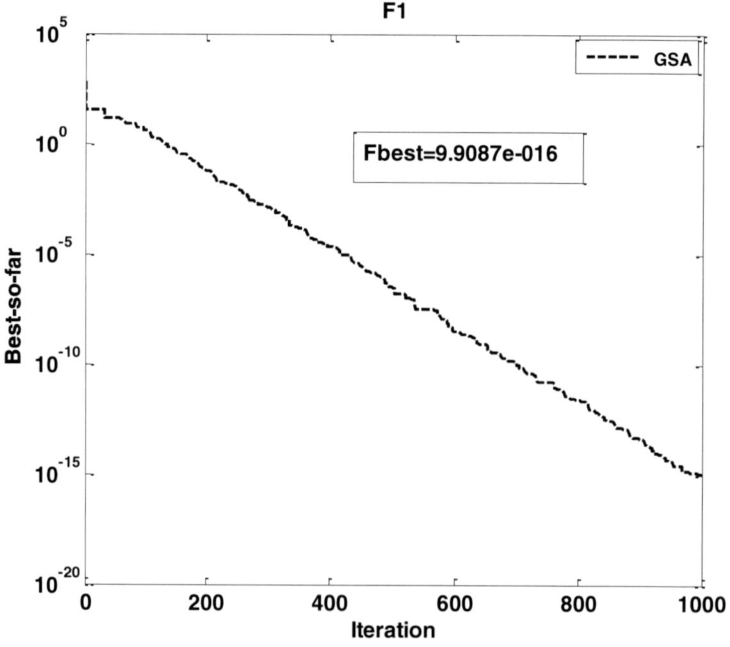

Figure 10. Convergence characteristic of GSA for minimization of F1 with n = 30.

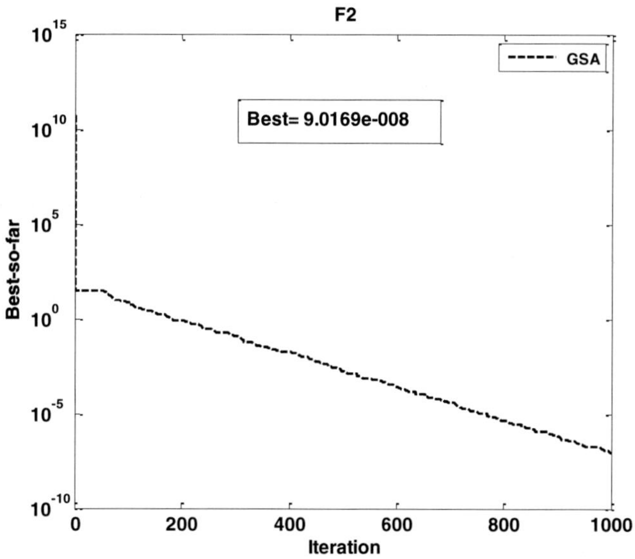

Figure 11. Convergence characteristic of GSA for minimization of F2 with n = 30.

Test 2: Dynamic Economic Dispatch based GSA

Case 1: 5 Unit System

In this case the five unit system data such as cost coefficient, B_{ij} loss coefficient, ramp rate limits, power generation limits and load demand are taken from [24-28] and given in the Appendix. In this test case transmission losses are considered, valve point effect and ramp rate limits are also taken in consideration. Figure 12 shows the characteristic of load demand for 24 hour. Table 2 shows the obtained results for this system, the best total production cost using the proposed method is 43223.78 $. All practical constraints of units are within the security limits. Results obtained using this technique are compared with a variety of recent optimization techniques such as: Simulated annealing (SA), adaptive particle swarm optimization (APSO) cited in [28], and artificial immune system (AIS) [28]. It can be clearly observed that the obtained results with GSA are less than those of reported in literature. Figure 13 shows convergence cost characteristic for one hour duration (PD=410 MW). Comparison between different methods is given in Table 3.

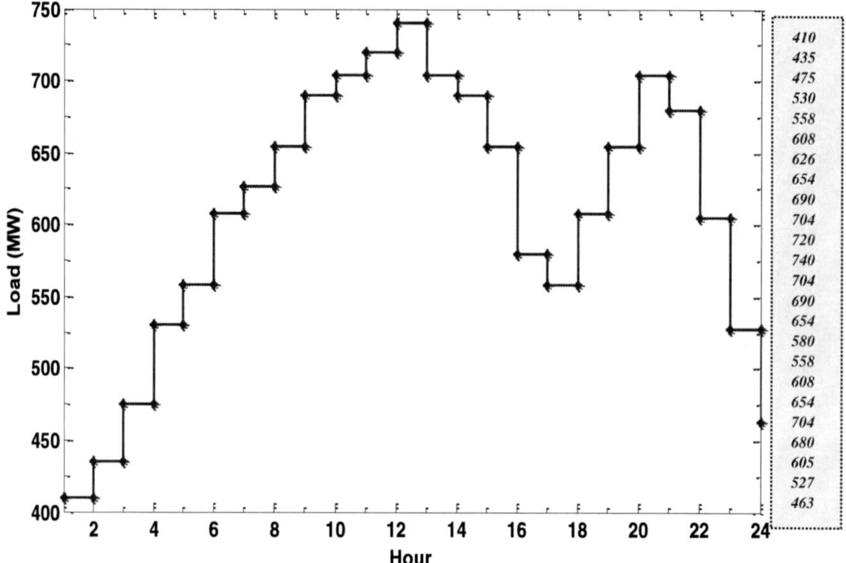

Figure 12. Hourly load profile (in MW) for 5-unit test systems.

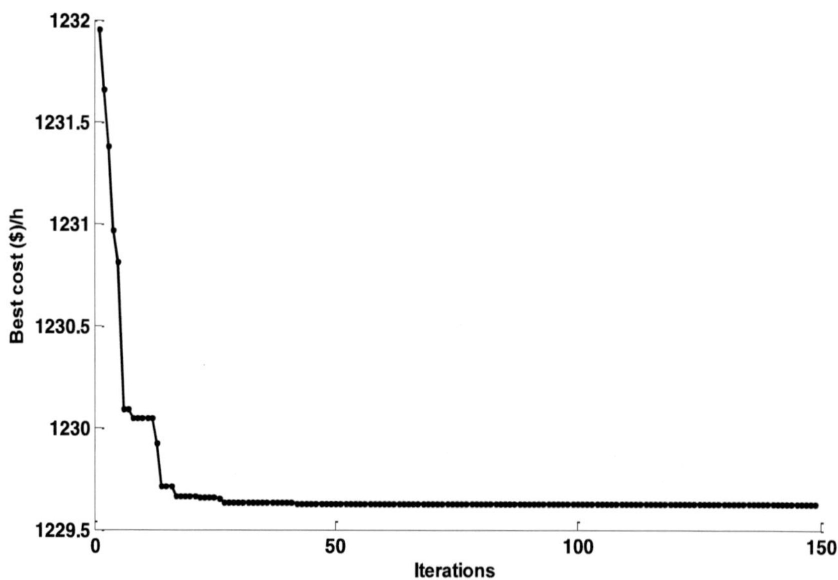

Figure 13. Convergence cost characteristic for one hour duration (PD=410 MW): 5 generating units.

Table 2. Generation shudle using Gravitational Search Algorithm (GSA): Case 1:5 unit considering loss

Hour	**P1**	**P2**	**P3**	**P4**	**P5**	Cost ($)	Ploss (MW)
1	10.2987	20.0000	30.00000	124.3350	229.3540	1229.8	3.98770
2	19.3766	20.0000	30.00000	140.6960	229.3700	1419.7	4.44260
3	10.2989	20.0000	30.00000	190.6960	229.3690	1496.6	5.36390
4	10.4477	20.0000	66.87300	209.6660	229.3700	1666.5	6.35670
5	10.4483	20.0000	95.36100	209.6660	229.3650	1672.0	6.84030
6	14.5462	49.8500	112.5250	209.6660	229.3700	1831.6	7.95720
7	10.5969	72.3010	112.5230	209.6660	229.3700	1846.4	8.45690
8	13.8025	97.8900	112.5240	209.6650	229.3700	1808.1	9.25150
9	43.3075	102.630	115.2030	209.6670	229.3700	2014.6	10.1775
10	64.6296	98.3900	112.5210	209.6490	229.3690	1999.2	10.5586
11	74.9999	99.2380	117.7315	209.6669	229.3699	2043.6	11.0062
12	74.9997	125.000	112.6770	209.6700	229.3735	2182.9	11.7202
13	64.6097	98.3900	112.5230	209.6660	229.3700	1999.1	10.5587
14	50.2189	98.3900	112.5230	209.6660	229.3690	1981.4	10.1669
15	36.4903	98.3900	112.5230	186.3500	229.3700	2015.3	9.12330
16	10.5968	98.3900	112.5240	136.3500	229.3700	1687.1	7.23080
17	10.5984	87.4360	112.5220	124.7550	229.3690	1621.4	6.68040
18	10.5963	98.3900	112.5240	165.0680	229.3700	1858.5	7.94830
19	13.8016	97.8900	112.5240	209.6660	229.3700	1808.1	9.25160
20	43.3063	119.789	112.5240	209.6660	229.3700	2118.5	10.6553
21	39.9498	98.3900	112.5240	209.6660	229.3700	1949.3	9.89980
22	10.5969	98.3910	110.0540	164.4690	229.3700	1865.9	7.88090
23	10.5985	98.3900	70.05400	124.7580	229.3700	1648.3	6.17050
24	10.5991	73.2160	30.05400	124.7580	229.3690	1459.7	4.99610
			Total Cost ($/h)	**43223.78**			

Table 3. Comparison of best results for Case 1: 5 units test System

Methods [28]	Best Cost($)-24 Hour	Average cost($)	Maximum cost($)
APSO	47356.00	NA	NA
AIS	44385.00	44758.8363	45553.7707
SA	44862.42	44921.7600	45893.9500
Proposed approach	**43223.78**	**43503.41**	**43870.14**

Case 2: 10 Unit System without Loss

In this second test case the 10 unit system with valve point loading effect is considered and the data can be obtained from [24-28]. The system transmission loss is neglected and the load demand is considered over 24 h time interval, Figure 14 shows Hourly load profile (in MW) for 10-unit test systems.

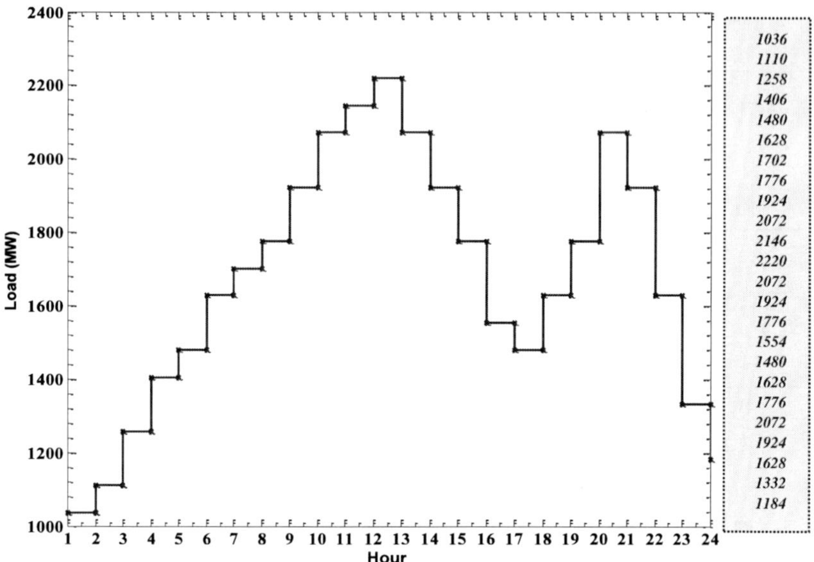

Figure 14. Hourly load profile (in MW) for 10-unit test systems.

Table 4. Comparison of best results for Case 2: 10 Generating units test System

Methods [28]	Best Cost($)- 24 Hour	Average cost($)	Maximum cost($)
EP-SQP	1031746.00	1035748.00	NA
PSO-SQP	1027334.00	1028546.00	1033986.00
MHEP-SQP	1028924.00	1031179.00	NA
IPSO	1023807.00	1023807.00	1026863.00
MDE	1031612.00	1033630.00	NA
AIS	1021980.00	NA	NA
HHS	1019091.00	NA	NA
Proposed approach	**1018766.25**		

The obtained optimal results are compared with results optimization techniques cited in the literature such as hybrid EP and SQP [28], Hybrid PSO-SQP [4], modified hybrid EP-SQP (MHEP-SQP) [6], improved PSO (IPSO) [10], modified differential evolution (MDE) [12], artificial immune system (AIS) [28], hybrid swarm intelligence based harmony search Algorithm (HHS) [9]. Figure 15 shows convergence cost characteristic for one hour duration (PD=1036MW). The detailed comparison for quality solutions (minimum, average and maximum costs) are presented in Table 4. It is evident from Table 4 that the proposed approach based GSA yields more optimal solution.

Test 3: Combined Economic-emission Dispatch with Smooth Cost Function

To demonstrate the performance of the proposed APSO, the algorithm is applied and tested to the 6 generating units without considering power transmission loss. The problem is solved as single and multi objective optimization. Detail data of this test system is given in [36].

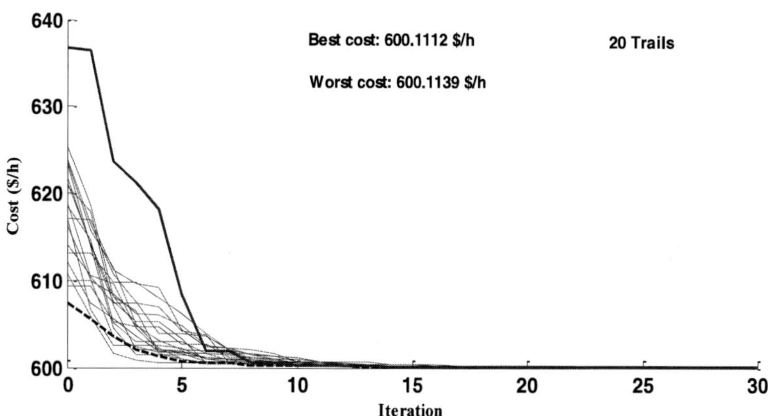

Figure 15. Convergence characteristic for optimal fuel cost based APSO.

Figure 15 demonstrates the robustness of the proposed algorithm, for 20 runs, the best cost is 600.1112 ($/h) and the worst cost is 600.1139 ($/h), the algorithm converges at low number of iteration. Figure 16-17 shows the convergence characteristics of the proposed APSO against the basic PSO.

Figure 18 shows the Convergence characteristic for combined fuel cost and emission. Table 5 summarizes the best solutions obtained for three cases.

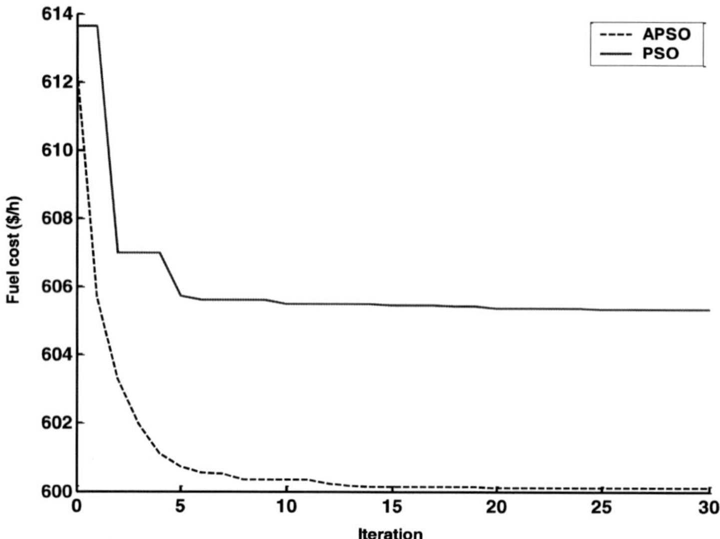

Figure 16. Convergence characteristic for optimal fuel cost: PSO, APSO.

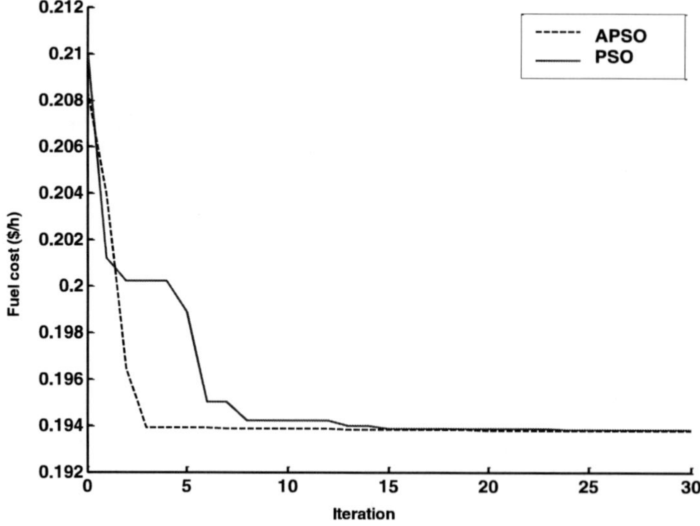

Figure 17. Convergence characteristic for optimal emission: PSO, APSO.

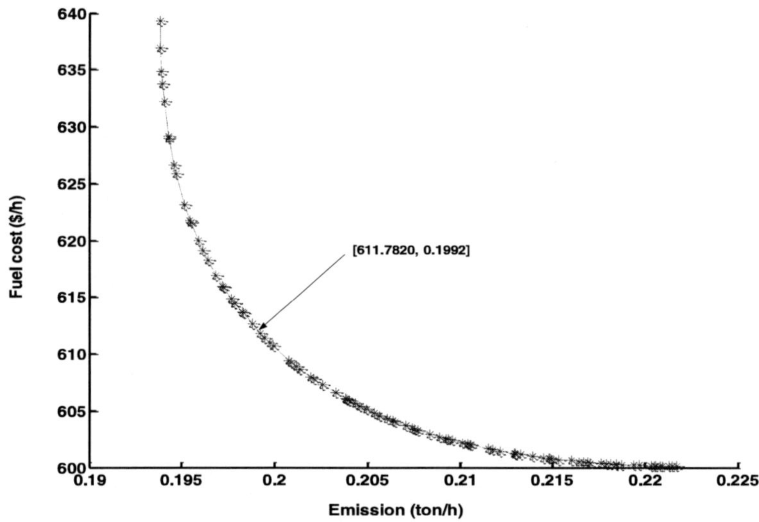

Figure 18. Convergence characteristic for combined fuel cost and emission.

Table 5. Control variables optimized using APSO

Control Variables (MW)	**Case 1** Fuel cost	**Case 2** Emissions	**Case3** Combined: Cost-Emissions
P_{G1}	10.9361	40.1295	10.4750
P_{G2}	29.9992	45.5277	29.8308
P_{G5}	52.4590	53.8264	52.7759
P_{G8}	101.6305	37.6679	101.6248
P_{G11}	52.3931	53.8113	52.7165
P_{G13}	35.9820	52.4372	35.9771
FC ($/h)	**600.1112**	**638.7807**	**611.8445**
Emission (ton/h)	**0.2219**	**0.1938**	**0.1992**
Loss (MW)	**0**	**0**	**0**

Test 4: Security Optimal Power Flow Considering Shunt FACTS

A standard IEEE 57-bus test system is considered, details data can be retrieved from Matpower [37], the total load demand is 1250.8 MW, the maximum and minimum values for voltages of all generating units and tap setting transformer control variables are considered to be 1.1-0.9 in p.u, the maximum and minimum values for voltages at all load buses are 1.06 and 0.94 in p.u, respectively.

Table 6. Control variables optimized: Test 4: case: IEEE 57-Bus

Control Variables(p.u)	APSO	Control Variables(p.u)	APSO
P_{G1}	1.432100	$T_{24\text{-}25}$	1.0500
P_{G2}	0.878400	$T_{24\text{-}25}$	1.0952
P_{G5}	0.449700	$T_{24\text{-}26}$	1.0154
P_{G6}	0.728800	$T_{7\text{-}29}$	1.0237
P_{G8}	4.611300	$T_{34\text{-}32}$	1.0028
P_{G9}	0.968100	$T_{11\text{-}41}$	1.0028
P_{G12}	3.599700	$T_{15\text{-}45}$	0.9450
V_{G1}	1.0567	T_{1446}	0.9765
V_{G2}	1.0547	$T_{10\text{-}51}$	0.9398
V_{G5}	1.0507	$T_{13\text{-}49}$	1.0059
V_{G6}	1.0698	$T_{11\text{-}43}$	1.0059
V_{G8}	1.0665	$T_{40\text{-}56}$	1.0290
V_{G9}	1.0547	$T_{39\text{-}57}$	0.9870
V_{G12}	1.0547	$T_{9\text{-}55}$	1.0500
$T_{4\text{-}18}$	1.0185	$Q_{sht}(18)$	0.2345
$T_{4\text{-}18}$	1.0269	$Q_{sht}(25)$	0.1384
$T_{21\text{-}20}$	1.0952	$Q_{sht}(46)$	-0.1172
$T_{24\text{-}25}$	1.0500	$Q_{sht}(53)$	0.1477
Fuel cost ($/h), **APSO**	**41713.8868**		
Ploss (MW)	16.009999		
Dv (p.u)	1.5827		
	Matpower [39]		
Fuel cost ($/h)	**41737.7900**		

The obtained optimal setting of control variables using the proposed APSO are given in Table 6. The total fuel cost achieved is 41713.8868 ($/h) which is less in comparison to the optimal value obtained from Matpower. Voltage magnitudes at all buses are within security (0.94-1.06 for load buses, and 0.9-1.1 for control buses) limits as well shown in Figure 19. Figure 20 shows the distribution of the angles values in degree at all buses. It is important to clarify that the attained results by the proposed approach satisfy the security constraints such as: reactive power for generators, the total load balance, and voltages limit at control buses (0.9-1.1) and load buses (0.94, 1.1).

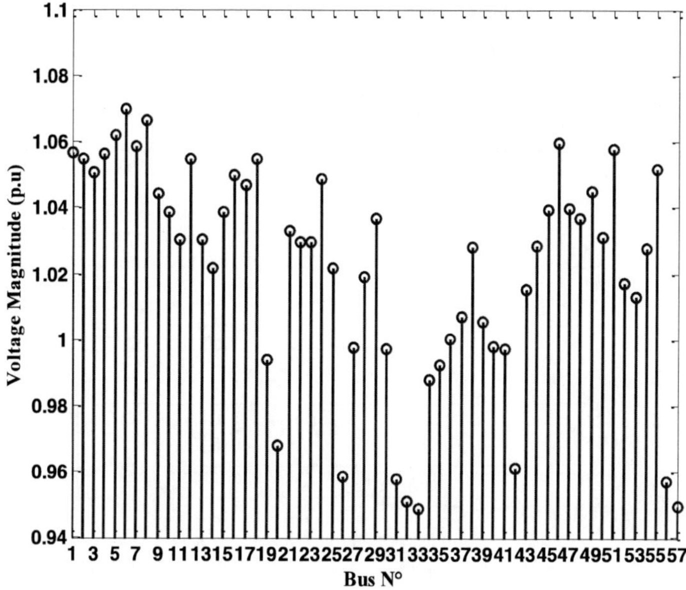

Figure 19. Distribution of voltage magnitudes at all buses.

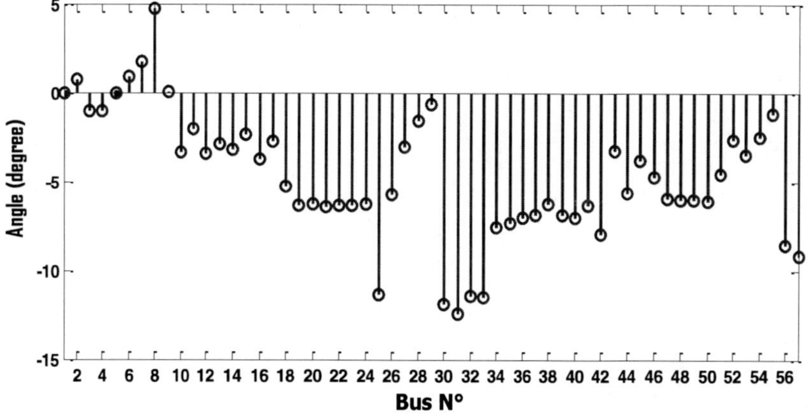

Figure 20. Distribution of angle at all buses.

Test 5. Economic Dispatch Considering Valve Point Effect

A system with 40 generators considering valve point loading effect is used here. Total load demand of the system is 10500 MW. This is a larger and complex system with many local optima. The system data can be retrieved from [17]. Table 7 shows economic dispatch results, a comparative study is given in Table 8, Figure 21 demonstrates the convergence characteristics of the proposed hierarchical APSO againt the standard PSO to solve large power system test considering practical generator constraints.

Table 7. Economic Dispatch Results for 40-Generating Units using the Proposed Approach: PD =10500 MW

Pg_i (MW)					
Unit N°	[DEBBO] [17]	[PDE]	Unit N°	[DEBBO] [17]	APSO
1	110.7998	110.7988	21	523.2794	523.2778
2	110.7998	110.8012	22	523.2794	523.2838
3	97.3999	97.3993	23	523.2794	523.2749
4	179.7331	179.7331	24	523.2794	523.2832
5	87.9576	87.7993	25	523.2794	523.3211
6	140.0000	140.0000	26	523.2794	523.2835
7	259.5997	259.5977	27	10.0	10.0000
8	284.5997	284.6003	28	10.0	10.0000
9	284.5997	284.5988	29	10.0	10.0000
10	130.0000	130.0000	30	97.0000	87.8000
11	168.7998	94.0000	31	190.0000	190.0000
12	94.0000	94.0000	32	190.0000	190.0000
13	214.7598	214.7594	33	190.0000	190.0000
14	394.2794	394.2803	34	164.7998	164.7998
15	394.2794	394.2794	35	200.0000	194.3594
16	304.5196	394.2787	36	200.0000	200.0000
17	489.2794	489.2752	37	110.0000	110.0000
18	489.2794	489.2783	38	110.0000	110.0000
19	511.2794	511.2783	39	110.0000	110.0000
20	511.2794	511.2831	40	511.2794	511.2753
TP (MW)				**10,500**	**10,500**
TC ($/h)				**121420.89**	**121413.3375**

Table 8. Comparison of the Proposed Approach with other Global Optimization Methods 40-Generating Units PD =10500 MW

Methods [21]	Minimum Cost($/h)
SA-PSO	121430.0000
SOH-PSO	121501.1400
DE	121416.2900
MPSO	122252.2600
PSO-SQP	122094.6700
NPSO	121704.7400
SOH-PSO	121501.1400
FAPSO	121712.4000
DEC-SQP	121741.9700
Our Approach	**121413.3375**

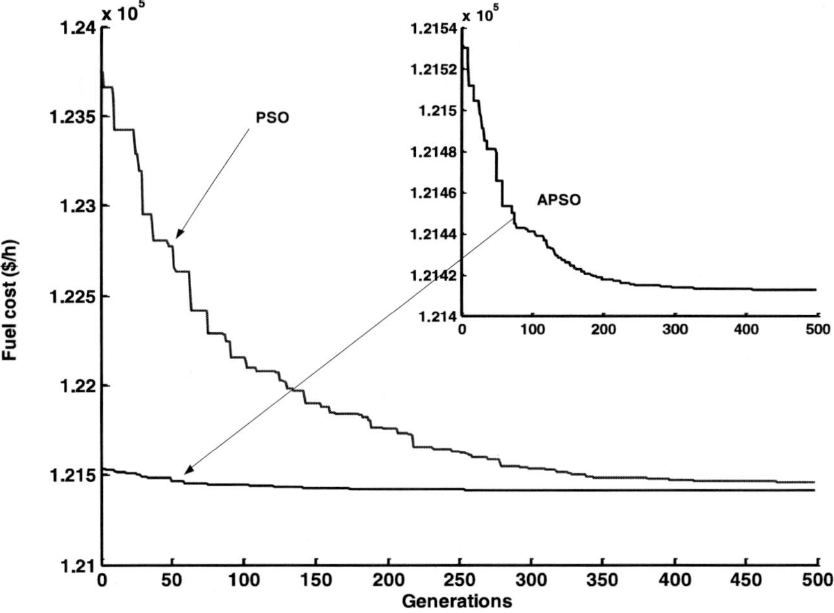

Figure 21. Convergence characteristic for 40 units with valve point effect.

APPENDIX A. TEST CASES' DATA

Table A.1. Generators cost coefficients for 10 unit test system

Unit	*Pmin [MW]*	*Pmax [MW]*	*UR*	*DR*	*a*	*b*	*c*	*e*	*f*
1	10	75	30	30	0.0080	2.0	25	100	0.0420
2	20	125	30	30	0.0030	1.8	60	140	0.0400
3	30	175	40	40	0.0012	2.1	100	160	0.0380
4	40	250	50	50	0.0010	2.0	120	180	0.0370
5	50	300	50	50	0.0015	1.8	40	200	0.0350

The B-matrix coefficients of the 5 unit test system in per unit in 100 MW base are presented as follows:

$$B_{ij} = \begin{pmatrix} 0.000049 & 0.000014 & 0.000015 & 0.000015 & 0.000020 \\ 0.000014 & 0.000045 & 0.000016 & 0.000020 & 0.000018 \\ 0.000015 & 0.000016 & 0.000039 & 0.000010 & 0.000012 \\ 0.000015 & 0.000020 & 0.000010 & 0.000040 & 0.000014 \\ 0.000020 & 0.000018 & 0.000012 & 0.000014 & 0.000035 \end{pmatrix}$$

Table A.2. Generators cost coefficients for 10 unit test system

Unit	*Pmin [MW]*	*Pmax [MW]*	*UR*	*DR*	*a*	*b*	*c*	*e*	*f*
1	150	470	80	80	0.00043	21.60	958.20	450	0.041
2	135	460	80	80	0.00063	21.05	1313.6	600	0.036
3	73	340	80	80	0.00039	20.81	604.97	320	0.028
4	60	300	50	50	0.00070	23.90	471.60	260	0.052
5	73	243	50	50	0.00079	21.62	480.29	280	0.063
6	57	160	50	50	0.00056	17.87	601.75	310	0.048
7	20	130	30	30	0.00211	16.51	502.70	300	0.086
8	47	120	30	30	0.00480	23.23	639.40	340	0.082
9	20	80	30	30	0.10908	19.58	455.60	270	0.098
10	55	55	30	30	0.00951	22.54	692.40	380	0.094

Table A.3. Generators cost coefficients for 40 unit test system

Bus Number	Pmin [MW]	Pmax [MW]	a	b	c	e	f
1	36	114	0.0069	6.73	94.705	100	0.084
2	36	114	0.0069	6.73	94.705	100	0.084
3	60	120	0.02028	7.07	309.54	100	0.084
4	80	190	0.00942	8.18	369.03	150	0.063
5	47	97	0.0114	5.35	148.89	120	0.077
6	68	140	0.01142	8.05	222.33	100	0.084
7	110	300	0.00357	8.03	278.71	200	0.042
8	135	300	0.00492	6.99	391.98	200	0.042
9	135	300	0.00573	6.6	455.76	200	0.042
10	130	300	0.00605	12.9	722.82	200	0.042
11	94	375	0.00515	12.9	635.2	200	0.042
12	94	375	0.00569	12.8	654.69	200	0.042
13	125	500	0.00421	12.5	913.4	300	0.035
14	125	500	0.00752	8.84	1760.4	300	0.035
15	125	500	0.00708	9.15	1728.3	300	0.035
16	125	500	0.00708	9.15	1728.3	300	0.035
17	220	500	0.00313	7.97	647.85	300	0.035
18	220	500	0.00313	7.95	649.69	300	0.035
19	242	550	0.00313	7.97	647.83	300	0.035
20	242	550	0.00313	7.97	647.81	300	0.035
21	254	550	0.00298	6.63	785.96	300	0.035
22	254	550	0.00298	6.63	785.96	300	0.035
23	254	550	0.00284	6.66	794.53	300	0.035
24	254	550	0.00284	6.66	794.53	300	0.035
25	254	550	0.00277	7.1	801.32	300	0.035
26	254	550	0.00277	7.1	801.32	300	0.035
27	10	150	0.52124	3.33	1055.1	120	0.077
28	10	150	0.52124	3.33	1055.1	120	0.077
29	10	150	0.52124	3.33	1055.1	120	0.077
30	47	97	0.0114	5.35	148.89	120	0.077
31	60	190	0.0016	6.43	222.92	150	0.063
32	60	190	0.0016	6.43	222.92	150	0.063
33	60	190	0.0016	6.43	222.92	150	0.063
34	90	200	0.0001	8.95	107.87	200	0.042
35	90	200	0.0001	8.62	116.58	200	0.042
36	90	200	0.0001	8.62	116.58	200	0.042
37	25	110	0.0161	5.88	307.45	80	0.098
38	25	110	0.0161	5.88	307.45	80	0.098
39	25	110	0.0161	5.88	307.45	80	0.098
40	242	550	0.00313	7.97	647.83	300	0.035

CONCLUSION

In this chapter two metaheuristic optimization methods have been proposed and applied for solving many practical problems related to power system operation and control. The first proposed method called gravitational search algorithm adapted to improve the solution of dynamic economic dispatch considering practical generator constraints (valve point effect and ramp rate limits). A new variant based PSO is proposed to solving multi objective optimal power flow considering distributed shunt FACTS technology, a dynamic mechanism search for PSO parameters in coordination with hierarchical decomposed strategy is proposed to enhance the performances of the original PSO method in term of solution quality and convergence characteristic.

The performance of the two proposed approaches has been tested with many practical electrical network, solving the dynamic economic dispatch to 5 and 10 units considering valve point effect, ramp rate limits and real power transmission losses, solving the combined economic emission dispatch on IEEE 30-Bus, solving the security OPF to standard IEEE 57-Bus with smooth cost function considering multi shunt FACTS devices, and applied to solving the static economic dispatch to large electrical network, 40 generating units considering valve point effect. The results of the proposed two algorithms compared with recent global optimization methods. It is observed that the proposed approaches are capable of finding the near global solution of non-linear and non-differentiable objective functions and obtain a competitive solution in term of solution quality and convergence charatceristic. Due to these efficient properties, in the future work, author will still to apply these two methods to solving the multi objective optimal power flow based multi FACTS devices considering practical generator units.

REFERENCES

[1] P. Pezzinia, O. Gomis-Bellmunt, and A. Sudrià-Andreua, "Optimization techniques to improve energy efficiency in power systems," *Renewable and Sustainable Energy Reviews*, pp.2029-2041, 2011.

[2] J. Wood and B. F. Wollenberg, Power Generation, Operation, and Control, 2nd Ed. New York: Wiley, 1984.

[3] J.-A. Momoh and J. Z. Zhu, "Improved interior point method for OPF problems," *IEEE Trans. Power* Systems, vol. 14, pp. 1114-1120, Aug. 1999.

[4] S. Frank, I. Steponavice, and S. Rebennak, "Optimal power flow: a bibliographic survey I, formulations and deterministic methods," *Int. J. Energy.System (Springer-Verlag),* 2012.

[5] F. Capitanescua, J. L. M. Ramos, P. Panciatici , D. Kirschend, A. M. Marcolini, L. Platbrood, and L. Wehenkel, "State-of-the-art, challenges, and future trends in security constrained optimal power flow, "*Int. J. Electr. Power System Res*, vol. 81, pp. 1731-1741, 2011.

[6] C.-L. Chiang, "Improved genetic algorithm for power economic dispatch of units with valve-point effects and multiple fuels," *IEEE Trans. Power Syst.*, vol. 20, no. 4, pp. 1690–1699, Nov. 2005.

[7] S. Hui, "Multi-objective optimization for hydraulic hybrid vehicle based on adaptive simulated annealing genetic algorithm," *Engineering Applications of Artificial Intelligence* 2010; 23(1):27e33.B.

[8] S. Frank, I. Steponavice, and S. Rebennak, "Optimal power flow: a bibliographic survey II, non-deterministic and hybrid methods," *Int. J. Energy.System (Springer-Verlag),* 2012.

[9] Z. L. Gaing, "Particle swarm optimization to solving the economic dispatch considering the generator constraints," *IEEE Trans. Power Systems*, vol. 18, no. 3, pp. 1187-1195, 2003.

[10] M. F. Mustafar, I. Musirin, M. R. Kalil, M. K. Idris,"Ant colony optimization (aco) based technique for voltage control and loss minimization using transformer tap setting," *In: Proc. 5th student conference on research and development SCOReD* 2007. 2007. p. 1–6.

[11] K. Price, R. Storn, and J. Lampinen, Differential Evolution: A Practical Approach to Global Optimization. Berlin, Germany: Springer- Verlag, 2005.

[12] N. Amjady and H. Sharifzadeh, "Security constrained optimal power flow considering detailed generator model by a new robust differential evolution algorithm," *Int. J. Electr. Power System Res*, vol. 81, pp. 740-749, 2011.

[13] Taher Niknam, "A new fuzzy adaptive hybrid particle swarm optimization algorithm for non-linear, non-smooth and non-convex economic dispatch," *Journal of Applied Energy,* Vol. 87, pp. 327-339, 2010.

[14] L. S. Coelho, R. C. Thom Souza, and V. Cocco Mariani, "Improved differential evoluation approach based on clutural algorithm and

diversity measure applied to solve economic load dispatch problems," *Journal of Mathemtics and Computers in Simulation*, Elsevier, 2009.

[15] S. Sivasubramani, K. S., Swarup, "Environmental economic dispatch using multi-objective harmony search algorithm," *Int. J. Electr. Power System Res*, vol. 81, pp. 1778-1785, 2011.

[16] P. K. Roy, S. P. Ghoshal, and S. S. Thakur, "Biogeography based optimization for multi-constraint optimal power flow with emission and non-smooth cost function," *International Journal of Expert Systems with Apllications*, vol. 37, pp. 8221-8228, 2010.

[17] A. Bhattacharya, and P. k, Chattopadhyay, "Hybrid differential evolution with Biogeaography-based optimization for solution of economic load dispatch," *IEEE Trans. Power Syst.*, vol. 25, no. 4, pp. 1955–1964, 2010.

[18] T. Niknam, H. D. Mojarrad, H. Z. Meymand, B. B. Firouzi, "A new honey bee mating optimization algorithm for non-smooth economic dispatch," *International Journal of Energy*, vol. 36, pp. 896-908, 2011.

[19] D. Karaboga, B. Basturk, A powerful and efficient algorithm for numeric optimization: artificial bee colony (ABC) algorithm, J. Global Optim. 39 (3) (2007) 459–471.

[20] B. Mahdad, T. Bouktir, K. Srairi, "OPF with Environmental Constraints with SVC Controller using Decomposed Parallel GA: Application to the Algerian Network," *Journal of Electrical Engineering & Technology*, Korea, vol. 4, No.1, pp. 55~65, March 2009.

[21] B. Mahdad, K. Srairi "Differential evolution based dynamic decomposed strategy for solution of large practical economic dispatch," *10th EEEIC International Conference on Environment and Electrical Engineering*, IEEE, Italy, 2011.

[22] Ke. Meng, H. G. Wang, Z. Y. Dong, and K. P. Wong, " Quantum-inspired particle swarm optimization for valve-point economic load dispatch," *IEEE Trans. Power Systems,* vol. 25, no. 1, pp. 215-222, 2010.

[23] Victoire, T., & Jeyakumar, A. E. (2005a), "Deterministically guided PSO for dynamic dispatch considering valve-point effect," *Electric Power Systems Research*, 73(3), 313–322.

[24] Victoire, T., & Jeyakumar, A. E. (2005b), "A modified hybrid EP–SQP approach for dynamic dispatch with valve-point effect," *Electrical Power Energy Systems*, 27(8), 594–601.

[25] R. Kumar, D. Sharma, and A. Sadu, "A hybrid multiagent based particle swarm optimization algorithm for economic power dispatch," *Int Journal of Elect. Power Energy Syst.*, vol. 33, pp. 115–123, 2011.

[26] A. I. Selvakumar, "Enhanced cross-entropy method for dynamic economic dispatch with valve-point effects," *Electrical Power and Energy Systems*, vol. 33, pp. 783–790, 2011.

[27] J.-C. Lee, W.-M. Lin, G.-C. Liao and T.-P. Tsao, "Quantum genetic Algorithm for dynamic economic dispatch with valve-point effects and including wind power system," *Electrical Power and Energy Systems*, vol. 33, pp. 189–197, 2011.

[28] S. Hemamalini and S. P. Simon, "Dynamic economic dispatch using artificial immune system for units with valve-point effect," *Electrical Power and Energy Systems*, vol. 33, pp. 868–874, 2011.

[29] B. Panigrahi, P. V. Ravikumar, and D. Sanjoy, "Adaptive particle swarm optimization approach for static and dynamic economic load dispatch," *Energy Conversion and Management*, vol. 49, pp. 1407–1415, 2008.

[30] X. Yuan, A. Su, Y. Yuan, H. Nie, and L. Wang, "An improved pso for dynamic load dispatch of generators with valve-point effects," *Energy*, vol. 34, pp. 67–74, 2009. August 18, 2011.

[31] Z. J. Wang, Ying, H. Qin, and Y. Lu, "Improved chaotic particle swarm optimization Algorithm for dynamic economic dispatch problem with valve-point effects, " *Energy Conversion and Management*, vol. 51, pp. 2893 – 2900, 2010.

[32] Y. Lu, J. Zhou, H. Qin, Y. Li, and Y. Zhang, "An adaptive hybrid differential evolution Algorithm for dynamic economic dispatch with valve-point effects," *Expert Systems with Applications,* vol. 37, pp. 4842– 4849, 2010.

[33] E. Atashpaz-Gargari and C. Lucas, "Imperialist competitive Algorithm: An Algorithm for optimization inspired by imperialistic competition," *in Evolutionary Computation*, 2007. CEC 2007. *IEEE Congress on,* 2007, pp. 4661– 4667.

[34] E. Rashedi, H. Nezamabadi-pour, S. Saryazdi, 'GSA: A Gravitational Search Algorithm,' *Information Sciences Journal*, vol. 179, pp. 2232-2248, 2009.

[35] B. Mahdad and K. Srairi, "Gravitational Search Algorithm to Solving Practical Dynamic Economic Dispatch," *Accepted at The International Conference on the European Energy Market*, EEM12, Italy, 2012.

[36] S. Duman, U. Guvenc, Y. Sonmez, N. Yorukeren, "Optimal power flow using gravitational Search Algorithm," *Energy Conversion and Management*, vol. 59, pp. 86-95, 2012.

[37] B. Mahdad, "Adaptive Fuzzy Controlled PSO Based decomposed network strategy to solve large scale economic dispatch,"*Chapter published in a Book (Advances in Energy Research. Volume 7):* Editors: Morena J. Acosta, Nova Publisher, USA, 2011.

[38] B. Mahdad, and K. Srairi, "Hierarchical Adaptive PSO for Multi-Objective OPF Considering Emissions Based Shunt FACTS," *Accepted at The International Conference IEEE, IECON12*, Canada, 2012.

[39] M. A. Abido, "Environmental/economic power dispartch using multiobjective evoulutionary algorithm," *IEEE Trans. Power Systems*, vol. 18, no. 4, pp. 1529-1537, May 2002.

[40] www.pserc.cornell.edu/matpower/, *Package of MATLAB, MATPOWER*.

In: Fuzzy Logic
Editor: Dinko Vukadinovic ISBN: 978-1-62417-151-2

Chapter 4

The Possibility of Applying Fuzzy Logic Outside the Range of E-Technologies

Djurdjica Parac-Osterman*[*], *Martinia Ira Glogar* and *Tomislav Rolich
University of Zagreb, Faculty of Textile Technology,
The Republic of Croatia

Abstract

A significant experimental work has been reported in literature considering applying fuzzy logic reasoning to variety of industrial processes. In this chapter some of the most interesting examples of fuzzy logic usage in colour matching processes will be presented, also interesting cases of fuzzy logic usage in textile dyeing process will be reviewed, and finally the part of a research work of surface structure influences on coloured textiles using fuzzy logic reasoning and approach, performed by authors, will be presented. As possibility of using fuzzy logic approach in colour matching processes on coloured textile, we will present the original scientific research of surface structure influences on coloured textiles applying fuzzy logic. As textile is a highly heterogeneous in its structure, it is often a difficult task to achieve the satisfactory colour similarity on the different structured textile materials.

[*] martinia.glogar@ttf.hr

The modern approach to improved colour matching is through application of fuzzy logic system in aim of achieving the advanced way of observing the influences of surface structure to colour similarity. It is a rule – based approach which is tolerant of imprecise data and any set of input-output data can be created as a fuzzy system. In this chapter the fuzzy logic based approach was performed in studying the influence of a different textile surface structure on colour appearance and similarity. For the analyses 100% cotton samples of certain characteristics were chosen. In the step considering fuzzy logic, the system was created based on terms that would define the surface characteristics of samples. The most important part was proper definition of rules and linguistic variables that use words for assigning values of properties that influence the experience of an observer (gloss, smoothness, roughness, etc.) and connect input space with output space. In this chapter the approach are based exactly on surface – structural characteristics of textile samples and the main idea was to examine the approach using fuzzy logic based technique in order to provide the method that would include surface parameters of textile samples that are of importance in colour applying on a different structured textile surfaces. It was confirmed that the fuzzy logic reasoning would have its application in control of surface structure influence of coloured textiles, especially for lighter shades where the influence of structure parameters on the colour experience of an observer is more emphasized.

1. INTRODUCTION

Fuzzy logic has been very successfully applied to many industrial applications over the past few decades [1-4].

To understand the reasons for the growing use of fuzzy logic it is necessary, first, to clarify what is meant by fuzzy logic. Fuzzy logic has two different meanings. In a narrow sense, fuzzy logic is a logical system, which is an extension of multi – valued logic. But in a wider sense, which is in predominant use today, fuzzy logic is almost synonymous with the theory of fuzzy sets, a theory which relates to classes of objects with un-sharp boundaries in which membership is a matter of degree. What is important to recognize is that, even in its narrow sense, the agenda of fuzzy logic is very different both in spirit and substance from the agendas of traditional multi – valued logical systems [5].

Fuzzy logic poses the ability to mimic the human mind to effectively employ modes of reasoning that are approximate rather than exact. In traditional engineering and industry, decisions or actions are based on precision, certainty and vigour. With fuzzy logic, the specifications can be mapped by rules in terms of words rather than numbers. Computing with the words explores imprecision and tolerance. In most applications, a fuzzy logic solution is a translation of a human solution. Further, fuzzy logic can model nonlinear functions of arbitrary complexity to a desired degree of accuracy. Fuzzy logic is a convenient way to map an input space to an output space. Fuzzy logic is one of the tools used to model a multi-input, multi-output system. Fuzzy logic deals with issues such as forming impressions and reasoning on a semantic or linguistic level.

Fuzzy logic provides the computerizing of human reasoning and it provides the ability to handle control problems when there is uncertainty due to complex dynamics of an environment. The advantage of fuzzy logic usage lies in basic characteristics which defines the simplicity and applicability of fuzzy logic system. Fuzzy logic is easy to understand and mathematical concepts behind fuzzy reasoning are simple. It is tolerant of imprecise data and any set of input-output data can be created as a fuzzy system. Fuzzy logic is a method that interprets the values in the input vector and, based on some set of rules, assigns values to the output vector. It is a rule-based approach that is particularly suited to a complex system where accurate mathematical models cannot give the satisfactory performance [6-10].

Another basic concept in fuzzy logic, which plays a central role in most of its applications, is that of a fuzzy if-then rule or, simply, fuzzy rule. Although rule-based systems have a long history of use, what is missing in such systems is machinery for dealing with fuzzy consequents and/or fuzzy antecedents. In fuzzy logic, this machinery is provided by what is called the calculus of fuzzy rules. The calculus of fuzzy rules serves as a basis for what might be called the Fuzzy Dependency and Command Language (FDCL). In this connection, what is important to recognize is that in most of the applications of fuzzy logic, a fuzzy logic solution is in reality a translation of a human solution into FDCL [11].

In this chapter some of the most interesting examples of fuzzy logic usage in colour matching processes will be presented, also interesting cases of fuzzy logic usage in textile dyeing process will be reviewed, and finally the original research work of possibility of using fuzzy logic reasoning and approach in colour matching processes in textile, performed by authors, will be presented.

The interesting research performed in the field of textile technology is one by B. Smith and J. Lu [12]. In their researches they found that non parametric methods of control in textile dyeing process, which includes neural network approach or fuzzy logic approach based on expert system, are highly appropriate, due to the complexity and uncertainty of dyeing process which is very difficult to control because of the numerous parameters influencing the final result.

Other interesting research regarding colour matching is one performed by W. Cheetham and J. Graf [13]. In every colour matching process, a number of parameters including also the colorant selection influence the final colour match with customers colour standard. The authors in their research developed the technique that involved fuzzy logic to compare the quality of a colour match for each influencing parameter.

V. Bombardier, E. Schmitt and P. Charpetier developed a vision system according to a fuzzy based sensor concept with aim to improve the wooden board's colour matching [14].

S. Gorji Kandi and M. Amani Tehran reports research from the field of graphic technologies [15]. They introduce a novel initialization method for fuzzy c-mean algorithm for clustering colour images in graphic technologies.

As possibility of using fuzzy logic approach in coloured textile analyses, the authors will present the part of the original scientific research on surface structure influences on coloured textiles applying fuzzy logic. As textile is a highly heterogeneous in its structure, it is often a difficult task to achieve the satisfactory colour similarity on the different structured textile materials. In literature the number of reports about modelling the colour appearance of structured materials can be found. One of the most unconventional approaches was made by Allen and Goldfinger, presented by Tsoutseos, A. A. and Nobbs, J. H. in paper: Alternative Approach to Colour Appearance of Textile Materials with Application to the Wet/Dry Reflectance Prediction [16]. The modern approach to improved colour matching is through application of fuzzy logic system in aim of achieving the advanced way of observing the influences of surface structure to colour similarity. In this chapter the fuzzy logic based approach was performed in studying the influence of a different textile surface structure on colour appearance and similarity. For the analyses 100% cotton samples of certain characteristics were chosen. In the step considering fuzzy logic, the system was created based on terms that would define the surface characteristics of samples. The most important part was proper definition of rules and linguistic variables that use words for assigning values of properties that influence the experience of an observer (gloss, smoothness, roughness, etc.) and connect input space with output space. In this work the approach are based exactly on surface – structural characteristics of textile samples and the main idea was to examine the approach using fuzzy logic based technique in order to provide the method that would include surface parameters of textile samples that are of importance in colour applying on a different structured textile surfaces. It was confirmed that the fuzzy logic reasoning would have its application in control of surface structure influence of coloured textiles, especially for lighter shades where the influence of structure parameters on the colour experience of an observer is more emphasized.

2. Basics of Fuzzy Logic

Fuzzy logic was initiated in 1965 [17], [18], [19], by Lotfi A. Zadeh. Fuzzy logic (FL) is a multi -valued logic, which allows intermediate values to be defined between conventional evaluations like true/false, yes/no, high/low, etc. Notions like rather tall or very fast can be formulated mathematically and processed by computers, in order to apply a more human-like way of thinking in the programming of computers [20].

A fuzzy system is an alternative to traditional notions of membership set and logic that has its origins in ancient Greek philosophy. The precision of mathematics owes its success in large part to the efforts of Aristotle and the philosophers who preceded him. In their efforts to devise a concise theory of logic, and later mathematics, the so-called" Laws of Thought" were posited [21]. One of these, the "Law of the Excluded Middle", states that every proposition must either be true or false. Even when Parminedes proposed the first version of this law (around 400 B.C.) there were strong and immediate objections: for example, Heraclitus proposed that things could be simultaneously true and not true.

It was Plato who laid the foundation for what would become fuzzy logic, indicating that there was a third region (beyond true and false) where these opposites "tumbled about". Other, more modern philosophers echoed his sentiments, notably Hegel, Marx and Engels. But it was Lukasiewicz who first proposed a systematic alternative to the bi–valued logic of Aristotle [22]. Even in the present time some Greeks are still outstanding examples for fussiness and fuzziness [23]. Fuzzy logic has emerged as a profitable tool for the controlling and steering of systems and complex industrial processes, as well as for household and entertainment electronics, as well as for other expert systems and applications.

In general, fuzzy logic system is a nonlinear mapping of an input data (feature) vector into a scalar output (the vector output case decomposes into a collection of independent multi-input/single-output system). The mapping process involves input/output membership functions, FL operators, fuzzy if–then rules, aggregation of output sets, and defuzzification [23, 24, 25].

A general model of a fuzzy inference system (FIS) is shown in Figure 1. It can be seen from the figure that the FIS contains four components: the fuzzifier, inference engine, rule base and defuzzifier. The rule base contains linguistic rules that are provided by experts. It is also possible to extract rules from numeric data. Once the rules have been established, the FIS can be viewed as a system that maps an input vector to an output vector. The fuzzifier maps input numbers into corresponding fuzzy memberships. This is required in order to activate rules that are in terms of linguistic variables. The fuzzifier takes input values and determines the degree to which they belong to each of the fuzzy sets via membership functions. The inference engine defines mapping from input fuzzy sets into output fuzzy sets. It determines the degree to which the antecedent is satisfied for each rule. If the antecedent of a given rule has more than one clause, fuzzy operators are applied to obtain one number that represents the result of the antecedent for that rule. It is possible that one or more rules may fire at the same time. Outputs for all rules are then aggregated. During aggregation, fuzzy sets that represent the output of each rule are combined into a single fuzzy set. Fuzzy rules are fired in parallel, which is one of the important aspects of an FIS. In an FIS, the order in which rules are fired does not affect the output [23, 24, 25].

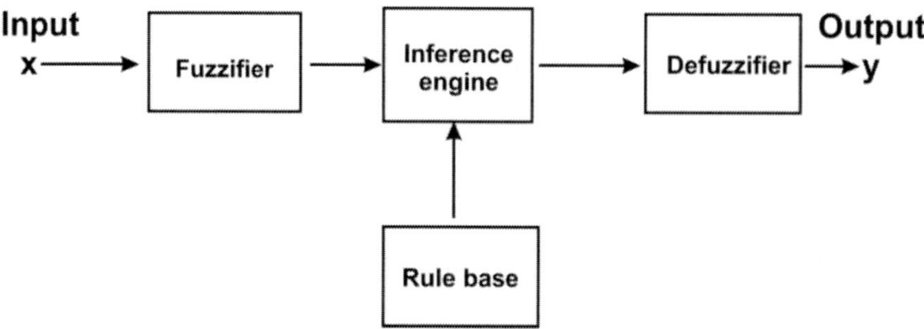

Figure 1. Block diagram of a fuzzy inference system

The defuzzifier maps output fuzzy sets into a crisp number. Given a fuzzy set that encompasses a range of output values, the defuzzifier returns one number, thereby moving from a fuzzy set to a crisp number. Several methods for defuzzification are used in practice, including the centroid, maximum, mean of maxima, height, and modified height defuzzifier. The most popular defuzzification method is the centroid, which calculates and returns the centre of gravity of the aggregated fuzzy set. FISs employ rules. However, unlike rules in conventional expert systems, a fuzzy rule localizes a region of space along the function surface instead of isolating a point on the surface. For a given input, more than one rule may fire. Also, in an FIS, multiple regions are combined in the output space to produce a composite region [23, 24, 25]. A general schematic of an FIS is shown in Figure 2.

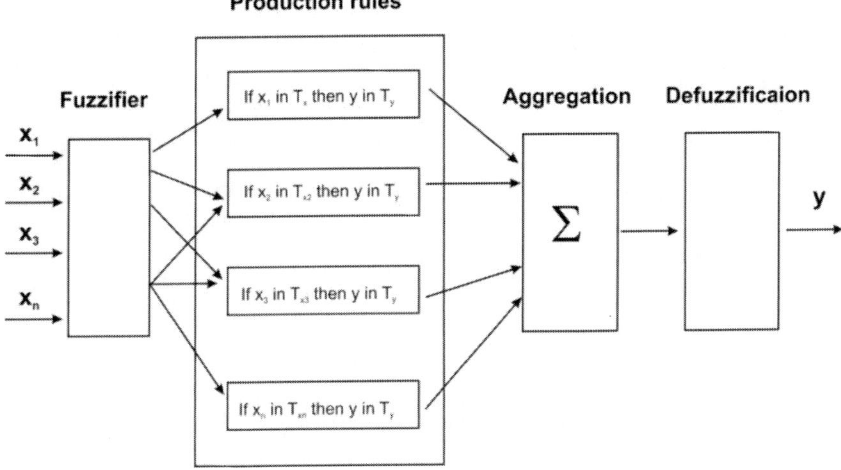

Figure 2. Schematic diagram of a fuzzy inference system.

The richness of fuzzy logic is that there are enormous numbers of possibilities that lead to lots of different mapping. This richness does require a careful understanding of fuzzy logic and the elements that comprise a fuzzy logic.

Fuzzy logic provides the computerizing of human reasoning and also provides the ability to handle control problems when there is uncertainty due to complex dynamics of an environment. The advantage of compactness of fuzzy set representation versus bivalent logic in production-rule systems lies in capability of fuzzy logic to represent both qualitative and quantitative information. Fuzzy systems are appropriate if sufficient expert knowledge about the process is available and provides a powerful framework for expert knowledge representation. Fuzzy system can be developed and used to control the complex processes where accurate mathematical models cannot give the satisfactory performance. Before a system that interprets the rules is build, all the terms and the adjectives that describe them have to be defined. Fuzzy sets and fuzzy operators are the subjects and verbs of fuzzy logic [24, 25].

Fuzzy Sets

Fuzzy logic starts with the concept of a fuzzy set. In classical mathematics we are familiar with what we call crisp sets. A *fuzzy set* is a set without a crisp, clearly defined boundary. It can contain elements with only a partial degree of membership. To understand what a fuzzy set is, first it must be considered what is meant by a *classical set*. A classical set is a container that wholly includes or wholly excludes any given element. But, where the realm of sharp edged yes-no logic stops being helpful, fuzzy reasoning becomes valuable [8, 24, 25].

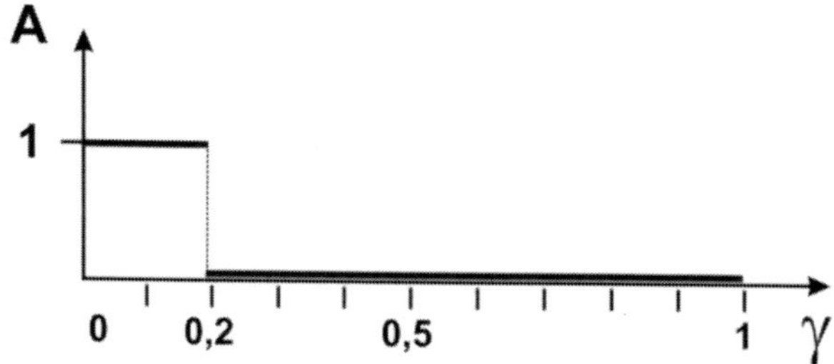

Figure 3. Characteristic function of a crisp set.

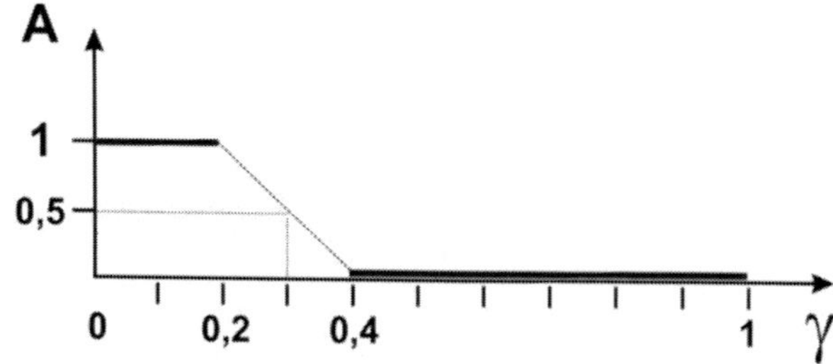

Figure 4. Characteristic function of a fuzzy set.

The fuzzy reasoning gives the ability to reply to a yes-no question with a not-quite-yes-or-no answer. This is the kind of thing that humans do all the time but it's a rather new trick for computers. Reasoning in fuzzy logic is just a matter of generalizing the familiar yes-no (Boolean) logic.

As the example on Figure 3 shows, for the classical mathematical, crisp set, the elements which have been assigned the number 1 can be interpreted as the elements that are in the set A and the elements which have assigned the number 0 as the elements that are definitively not in the set. So, if the "true" is given the numerical value of 1 than the "false" is given to the numerical value of 0, and that is definite [24, 25].

This concept is sufficient for many areas of applications, but it can easily be seen, that it lacks in flexibility for some applications. For such applications, a more natural way to construct the set would be to relax the strict separation between "true" and "false" (or any other linguistic value assigned for the purpose, for example "low" and "not low"). This can be done by allowing not only the crisp values, but also permitting values in-between "true" and "false", meaning 0 and 1, allowing values like 0.2 and 0.7453. Of course, again the number 1 assigned to an element means, that the element is in the set and 0 means that the element is definitely not in the set. All other values mean a gradual membership to the set (Figure 4). In fact, infinitely many alternatives can be allowed between the boundaries 0 and 1 [24, 25].

The element that represents the magnitude of participation of each input is a *membership function.* A *membership function* (MF) is a curve that defines how each point in the input space is mapped to a membership value (or degree of membership) between 0 and 1. It associates a weighting with each of the inputs that are processed, define functional overlap between inputs, and ultimately determines an output response. The rules use the input membership values as weighting factors to determine their influence on the fuzzy output sets of the final output conclusion [24, 25, 26, 27].

Membership Functions

Membership function is usually assigned as μ. In general, any membership function can be based on a shape of basic functions: linear function, Gaussian distribution curve, sigmoid function, square and cubed polynomial functions.

The simplest function shape is consists of partially straight segments. In a group of such functions the simplest is triangle membership function which consists of three points making the shape of triangle. Trapezoid function has a flat top. These straight line membership functions have the advantage of simplicity. The generalized bell membership function is defined with three parameters and consists of one parameter more than Gaussian distribution curve. Both of these curves have the advantage of being smooth and nonzero at all points, but they are unable to specify asymmetric membership functions, which are important in certain applications [24, 25, 26, 27].

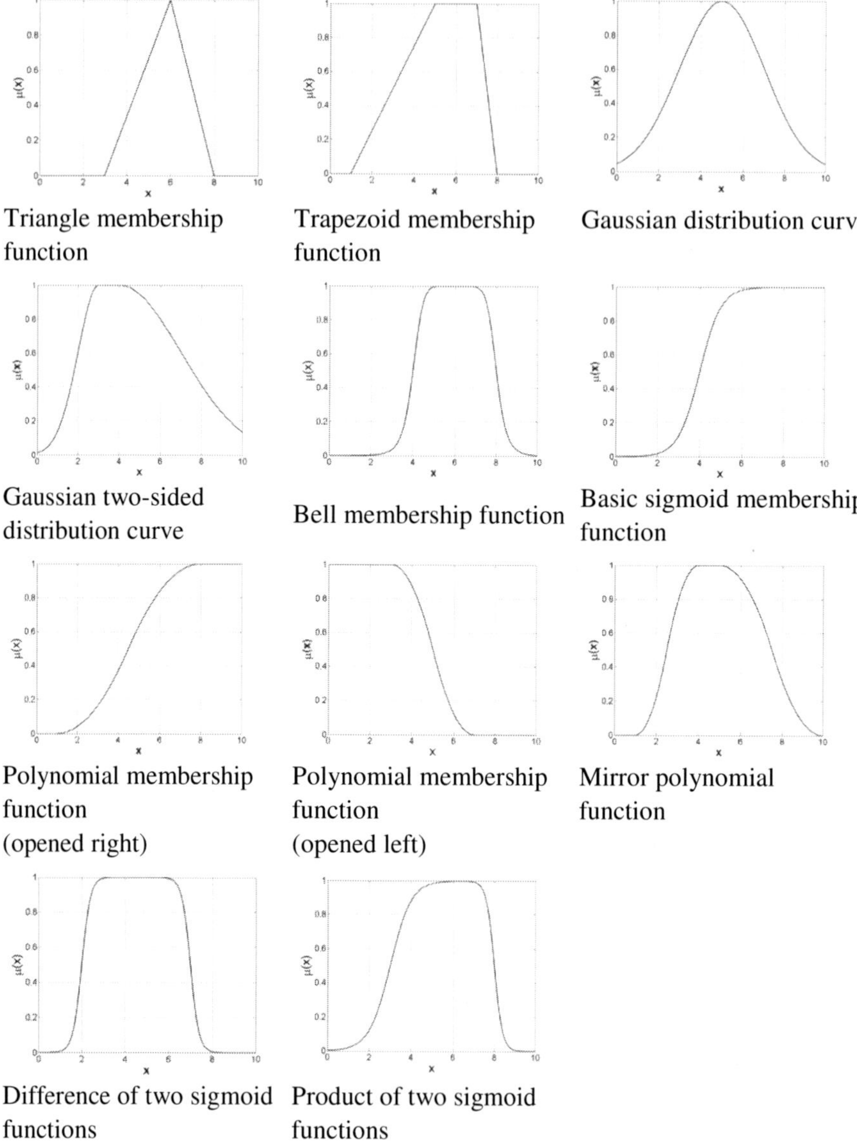

Figure 5. Shapes of the basics membership functions.

Sigmoid functions can be opened left or right. Asymmetric and closed membership functions can be synthesized using two sigmoid functions (as the difference between two sigmoid functions or product of two sigmoid functions).

Polynomial functions can be asymmetrically opened left, right, or closed on both sides.

Basic shapes of mostly used membership functions are shown on Figure 5, where the membership function is defined by ordinate, while values are defined on abscissa x. This is the way of defining the allowed values of certain linguistic variable [24, 25].

Linguistic Variables

Linguistic variable is a word or set of words (for example: density, thickness, gloss, relief, etc.), which have defined range of allowed linguistic values (for example: big, small, middle, positive, low, etc.), while only in some cases linguistic variable can be evaluated in numerical values (for example: thickness can be evaluated in numerical values but reliefness cannot).

The meaning of the linguistic variables is defined by the set, while the process of forming the set is defined with syntactic rule or generative grammar [24, 25, 26, 27].

The most usual generative grammar is grammar type 2 or, so called, grammar with no context. The most important parts of such grammar are sets of final terms or vocabulary, and set of production rules. Elements of vocabulary can be defined in several categories of which the most important are three:

a) *Primary terms*
(For example: big, small, middle, positive big, negative middle, fast, slow, unknown, ...),
b) *Fences*
(For example: highly, slightly, no, approximate, ...),
c) *Links*
(For example: and, or, but, ...).

With production rules, the allowed way of connecting the elements of vocabulary is defined, with aim to form the elements of primary terms set.

Linguistic models become deductive for computers only when the mathematical meaning is assigned to linguistic variables, using semantic rules which define the meaning of certain vocabulary categories [24, 25, 26, 27].

If – then Rules

These if-then rule statements are used to formulate the conditional statements that comprise fuzzy logic. A single fuzzy if-then rule assumes the form if *x* is *A* then *z* is *C* where *A* and *C* are linguistic values defined by fuzzy sets on the ranges (universes of discourse) X and Y, respectively. The if-part of the rule "*x* is *A*" is called the antecedent or premise, while the then-part of the rule "*z* is *C*" is called the consequent or conclusion.

The input for the implication process is a single number given by the antecedent, and the output is a fuzzy set. Implication occurs for each rule. If there are multiple parts to the antecedent, a fuzzy logic operators *AND*, *OR* and *NOT* has to be applied and than the if – then rule assimilate the following form: If *x* is *A AND y* is *B* then *z* is *C*; If *x* is *A OR y* is *B* then *z* is *C*; If *x* is *A AND y* is *NOT B* then *z* is *C*. The output of each rule is a fuzzy set. The output fuzzy sets for each rule are then aggregated into a single output fuzzy set. Finally the resulting set is defuzzified, or resolved to a single number [24, 25, 26, 27].

Fuzzy logic is expanded set of Boolean logic. Table of truth for *AND*, *OR* and *NOT* operators for Boolean logic have the following form (Table 1).

Table 1. Table of truth of Boolean logic

A	B	A AND B	A	B	A OR B	A	NOT A
0	0	0	0	0	0	0	1
0	1	0	0	1	1	1	0
1	0	0	1	0	1		
1	1	1	1	1	1		
AND			OR			NOT	

Graphically, such form of Boolean logic can be presented in following forms (Figure 6):

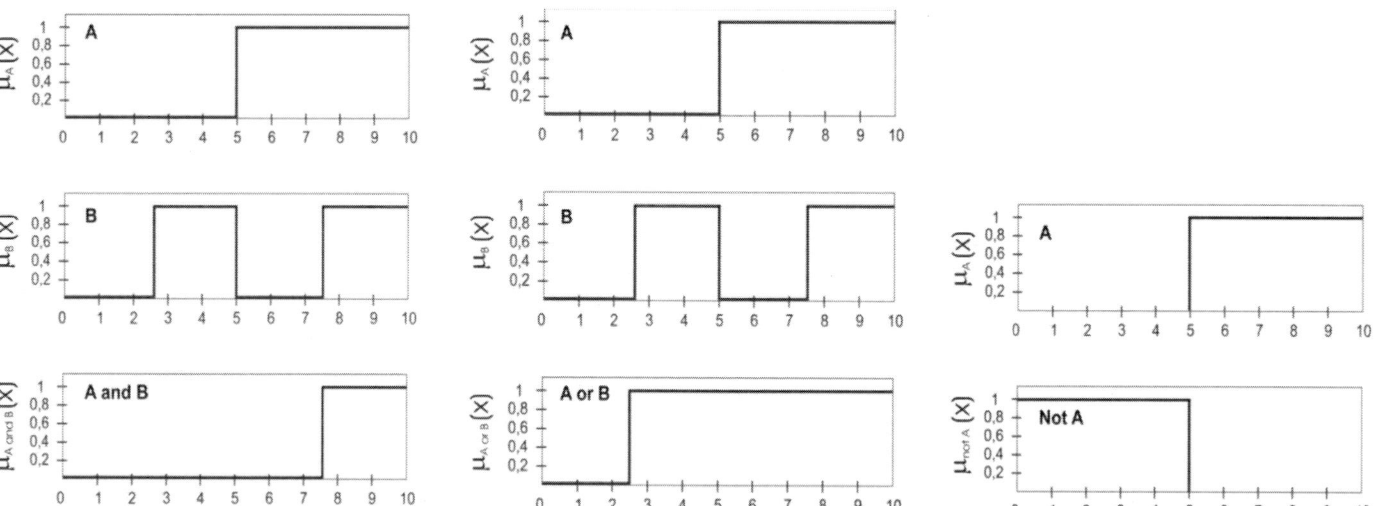

Figure 6. Graphic of Boolean logic.

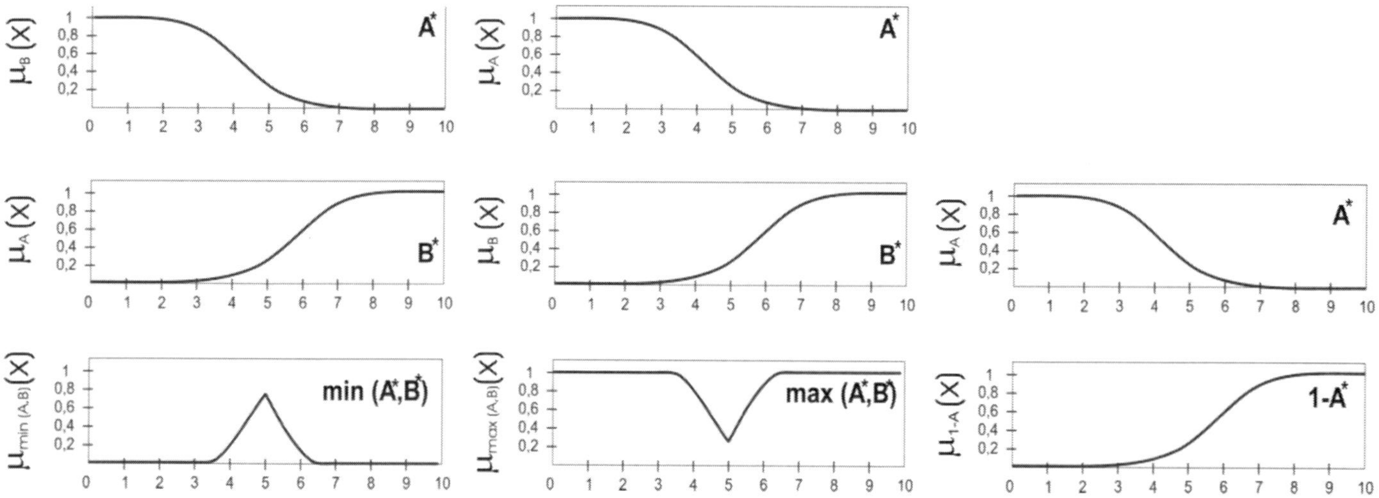

Figure 7. Membership functions related to table of truth in fuzzy logic.

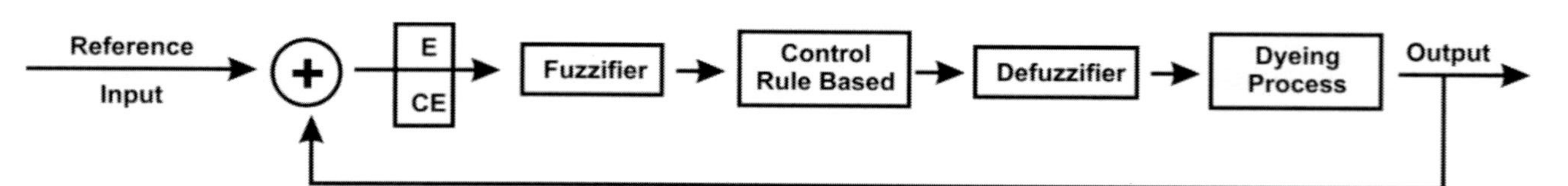

Figure 8. Fuzzy controller structure for dyeing process proposed by Smith and Lu.

In fuzzy logic the truth of any statement is gradient, any output value can be any real number between 0 and 1. So table of truth in fuzzy logic would have the following form (Table 2).

Table 2.Table of truth in fuzzy logic

*A**	*B**	*min (A*,B*)*	*A**	*B**	*max (A*,B*)*	*A**	*"not"1- A**
0	0	0	0	0	0	0	1
0	1	0	0	1	1	1	0
1	0	0	1	1	1		
1	1	1	1	1	1		
AND			*OR*			*NOT*	

Membership functions related to this table of truth can be as shown in Figure 7.

The basic agreements are shown on Figure 6 and Figure 7 between two – valued and multi – valued logic operators *AND*, *OR* and *NOT*. By operator *AND* the conjunction is defined, by operator *OR* separating is defined, and by operator "NOT" the fuzzy complement is defined. The most of the fuzzy logic based system in application are based on these three basic operators, although the literature suggests some additional operators like T binary operator or diverse T–norm and T–co-norm operators, which were suggested by Duboisea, Schweizera, Sklara or Sugena, in aim of achieving higher limits of functions or higher tolerances [24, 28, 29].

3. Fuzzy Logic in Industrial Application

The most usual application of fuzzy logic in industrial processes and engineering is in a form of fuzzy logic controllers. These have been very successfully applied to many industrial applications over the past few decades [30]. Initially outlined by Zadeh [18, 30] and explored by Mandami [31, 32] in early 1970s, fuzzy logic controllers applications exhibited their first industrial and commercial growth in Japan, just a decade later [33, 34]. Since then, many companies have offered consumer – oriented products enhanced by fuzzy logic controller technology.

A common feature of conventional control is that the control algorithm is analytically described by equations. In general, the synthesis of such control algorithms requires a formalized analytical description of the controlled system by a mathematical model [30, 35]. The main characteristic of fuzzy logic is its capability of expressing knowledge in a linguistic way, allowing a system to be described by simple, human friendly rules [30, 36, 37]. This alternative linguistic approach leads to a meaningful reduction of the computational burden. Usually fuzzy logic solutions are not aimed at achieving the computational precision of traditional techniques, but aim at finding acceptable solutions in a shorter time. The natural applications of fuzzy logic control (FLC) are those systems whose mathematical model is unknown, very complex or time-varying, and where human experience can play an important role.

As in other control applications, fuzzy logic implementation could be carried out either with general-purpose systems or dedicated systems, and many solutions have been proposed for both alternatives. The structure and relative simplicity of fuzzy processing algorithms naturally lead to straightforward implementation in dedicated hardware structures. However, the majority of fuzzy logic implementations reported in the literatures use general-purpose hardware due to the fact that such an implementation involves a low start-up cost with a well-defined control algorithm such as fuzzy logic controls [30].

Ying Bai, *Zvi. S. Roth* and *Hanqi Zhuang* [30] reported interesting research of applying fuzzy logic controller to suppress noises and coupling effects for a laser tracking system.

Implementation of robots in industries, manufacturing and other areas demand robots to perform movements with high accuracy. Calibration is a way to improve the accuracy of the performance of robots. Collecting a high quality measurement data for robot calibration is a challenging task in the robot calibration. Utilizing a laser tracking system (LTS) is a potential solution to handle this challenging problem [30].

The idea of using the laser interferometer transducer system as a robot measurement technique is developed by Lau, Hocken and Haight [38]. Others [30] introduced multi-laser interferometry for coordinate measuring. Harb [39] has reported a laser tracking system for multi-axis machine calibration. A 2D plane laser tracking measurement system is reported by Nakamura *et al.* [40]. A spherical seat bearing and a steel-ball rotor are utilized in this scheme. Rene *et al.* [41] have reported a portable laser tracking system using the triangulation principle. In order to improve the tracking and processing speed, some researchers [42] developed a new method to obtain the orientation information of the target on the end-effectors in real time. Vincze and Spiess *et al.* [43, 44] reported a laser tracking system to measure position and orientation of robot end-effectors under motion. Some other researchers have reported a laser tracking system [45] with a similar configuration as indicated in [43]. Its tracking speed, however, is relatively slow. Spiess *et al.* [30, 44] have addressed the calibration of a dynamic robot using a 6D laser tracking system.

Shieh and Li [46] developed an Integrated Fuzzy Logic Controller (IFLC) for a servomotor system. Kawaji *et al.* [47] developed a fuzzy model following a servo control system for a nonlinear DC servomotor. Betin *et al.* [48] developed a fuzzy logic that was applied to a speed control for a stepping motor drive to improve the tracking performance and robustness of the high stepping rates on stepping motor. Ha *et al.* [49] designed a robust control system using sliding-mode control that incorporates a fuzzy tuning technique. Li and Lee [50] developed and implemented a fuzzy motion control system. Chang *et al.* [51] reported a fuzzy P+ID controller for a mechanical manipulator. Rubaai *et al.* [52] designed and developed a microprocessor-based fuzzy logic tracking controller for motion controls and drives. The results showed excellent tracking performance for both speed and position trajectories.

Mario I. Chacón M [30] reported an interesting review of fuzzy logic application for image processing (definition and applications of a fuzzy image processing scheme).

Image processing is a prominent area that supports applications in different fields, such as medicine, astronomy, national security, autonomous systems, product quality, industrial applications, *etc.* [30, 53]. A high volume of research work is published every year in this area. One of the most challenge works is the development of image processing algorithms to try to simulate in some way the human visual system. As is known, these kinds of algorithms require modelling of complex systems, which require processing of information with high degree of uncertainty and subjectivity. A better structure to model a real system, in this case, the human visual system, must incorporate tools to model better subjective concepts and must be able to deal with the uncertainty of the information. A mathematical framework that incorporates these tools is fuzzy logic. In the area of image processing, especially in industrial applications, it is common to deal with subjective concepts like brightness, edges, uniformity, measurements, *etc.* So the incorporation of fuzzy logic into the development of image processing and analysis has opened a new research area in the image-processing field [30].

Guillermo Ayala, *Teresa León* and *Victoria Zapater* [30] presented a report on fuzzy logic for medical engineering, in more detail in an application to vessel segmentation. In their work they presented the problem of retinal vessel detection in fundus images. Optic fundus assessment has been widely used by the medical community for diagnosing different pathologies such as hypertension, diabetes, arteriosclerosis, cardiovascular disease and strokes. An automatic assessment for blood vessel anomalies in the optic fundus initially requires distinguishing the vessels from the background, which is the aim of the so-called segmentation methods [30].

Intelligent Transportation Systems (ITS) are systems utilizing multidisciplinary technologies among which are also fuzzy logic, to improve all kinds of transportation. One application of ITS is to provide assistance for controlling some vehicle components, like speed. This is known as the cruise control (CC) system. Such systems are already mass installed in top-of-the-line-end vehicles Adaptive cruise control (ACC) is a cruise control extension [30]

There are a lot of techniques for performing ACC. Conventional methods based on analytical control produce good results. Nevertheless, they have high design and computational costs since the application object, a car is a non-linear element and cannot be completely represented mathematically [53-56]. Other ways to achieve human-like speed control include the application of artificial intelligence techniques [30, 57]. One of these techniques, fuzzy control, allows approximate human reasoning and an intuitive control structure [30].

Fuzzy logic is a powerful, albeit somewhat controversial, technique. It permits control without extensive knowledge of the equations of the process and it very effectively represents human reasoning methods [58]. Nowadays, fuzzy logic is used to manage a wide range of systems, like cars, aircraft and railways [30]. The present application is for a car computer throttle control powered by a fuzzy logic controller, with the capability of performing adaptive cruise control in unmanned/manual driving.

In a road vehicle, the functions of transmission are to match the running state of the engine to the motion states of the vehicle. The aims are to optimize the fuel economy and the dynamic features, keep the vehicle safe and controllable, and make passengers more comfortable. The ratio of transmission, which is determined with the gear positions, is changed according to engine states, vehicle states, and road situations [30].

Transmissions are classified as Automatic Transmissions (AT) and Manual Transmissions (MT). A traditional AT is composed of a planetary gearbox and a torque converter. MT is composed of a gearbox and a dry clutch. It has to be operated manually. However, it has the advantages of high efficiency and is easy to manufacture and maintain. Automated Manual Transmission (AMT) achieves automatic shifting function by electronic control of MT. AMT has the advantages of both MT and AT [59, 60]. To implement automatic shifting AMT must have the ability of automatic gear ratio selection, *i.e.*, gear position decision (GPD), and the ability of automatic shifting control. Shifting control is a typical mechanical device control problem. Much research has been carried out to improve its performance [61, 62, 63].

With the development of the technology, some ideas have been presented from simple mechanical to software-based approaches. Among them are knowledge and fuzzy logic based tactics. Graf, Shimizu and Hayashi [30] presented neuro-fuzzy based GPD concepts.

Fuzzy Logic Control for automobiles also includes numerous researches in a field of navigation and collision avoidance system.

In 1990, *Nguyen and Widrow* [64] applied a self-learning neural network to the truck backer-upper problem. The latter has become an acknowledged benchmark in non-linear control since then and numerous other techniques have been tried, including genetic programming and neurogenetics, simplified neural network solution through problem decomposition [30], *etc.* Computational overhead is usually very high in such applications, *e.g.* in [30] it is shown that about 20,000 back-up cycles are needed before a neural network learns and, even then, the back propagation algorithm may not converge for some sets of training samples. A simplified version of the control problem (consisting of the cab part only), on the other hand, has been heavily exploited in the field of fuzzy control [65 – 69]. In these cases the traditional application area of fuzzy logic - knowledge-based control - would be an appropriate solution.

Also, very interesting and successful researches have been reported in a field of fuzzy logic based control mechanisms for handling the uncertainties facing mobile robots in changing unstructured environments. Autonomous mobile robots navigating in real-world changing and dynamic unstructured environments (like outdoor environments) need adequate control mechanisms to handle the uncertainties and imprecision present in such environments.

The fuzzy logic controller (FLC) is credited with being an adequate methodology for designing robust controllers that are able to deliver a satisfactory performance in face of uncertainty and imprecision [70]. The use of fuzzy rules and linguistic variables makes fuzzy control an adequate design tool for non-linear systems for which a precise mathematical model cannot be easily obtained, but for which heuristic control knowledge is available which is the case of mobile robots navigating in unstructured environments. As a result the FLC has become a popular approach to mobile robot control in recent years [70, 71, 72].

Fuzzy logic based control system found its application also in a field of flight control. The interesting are researches of usage of nonlinear optimal control of helicopter using fuzzy gain scheduling, and also fuzzy logic approach to path tracking and obstacles avoidance of unmanned aerial vehicles [30].

Helicopters exhibit high levels of agility and manoeuvrability, performing several actions, such as “climbing”, ”hovering”, and ”forward flight” [30]. A helicopter with superior performance is needed. With its high agility and performance, a helicopter is also unstable exhibiting non-linear qualities. In view of this, a more effective helicopter flight control strategy is developed to increase the stability and performance. Linear optimal controllers are designed in the flight envelope operating points. Linear quadratic (LQ) optimization has been utilized extensively in the field of aerospace engineering. LQ technology designs an optimal controller that minimizes a given performance index. A challenging task for control engineers is helicopter autopilot designs that achieve the desired robust stability and performance over a large flight envelope (defined as altitude *vs* Mach number). A typical control design method employed for helicopters operating over a large flight envelope is gain scheduling. There is no general procedure for the interpolation of the already available control law into a gain-scheduler. However, it is effective for gain-scheduler design using fuzzy logic [73]. The fuzzy inference system employing fuzzy if-then rules can control a plant using human knowledge. Fuzzy controllers have been widely used in industry because of their easy realization and robustness [30, 74]. Much work has been done on the analysis of control rules and membership function parameters. The Takagi-Sugeno (TS) fuzzy model can be regarded as a particular case of the general fuzzy model [30]. Two types of fuzzy controllers are combined to control an autonomous helicopter [75]. The Mamdani-type controller regulates the desired velocity and the TS-type controller achieves the desired attitude angle.

As far as the unmanned aerial vehicles (UAV) is concerns, these have attracted increasing attention in military and civilian applications, such as mapping, patrolling, search and rescue, and reconnaissance [30].

Recently, some methods of UAV/mobile vehicle navigation for obstacle avoidance have been reported [76, 77, 78]. However, most of the real situation is that obstacles might appear unexpectedly, and as such the pre-designed path might become “misleading” to the vehicle if no correcting action is taken.

Furthermore, obstacles may be of any shape (not necessarily a point mass) and may be either inert or hostile (including seeking a collision). How to avoid unexpected, irregular, even moving obstacles while maintaining close path tracking of UAV is a challenging problem. A fuzzy logic-based approach to path tracking and obstacle avoiding has been reported by Zhao Sun, Tao Dong, Xiaohong Liao, Ran Zhang and David Y. Song. The case that obstacles are either still or moving and appear along the pre-determined flight path unexpectedly has been considered. Fuzzy logic control algorithms are developed to achieve close path tracking while avoiding obstacles.

Due to the ability to handle unknown conditions, fuzzy logic system is an ideal tool to address the obstacle avoidance problem. In their work, *Zhao Sun*, *Tao Dong*, *Xiaohong Liao*, *Ran Zhang* and *David Y. Song* [30] combined model – based control with fuzzy logic control for UAV navigation in the presence of stationary or moving obstacles. This method provides a function for avoiding stationary and moving obstacles by sensing the distance and the angle between the UAV and the obstacles. Simulation studies on multiple obstacles with various shapes have been conducted and the effectiveness of the proposed method has been verified [30].

Since the human population is now living in an information age in which data are plentiful, the demands of information management, research, and analysis have become more and more important. Also, the amount of data generated and gathered has been increasing rapidly. This explosive growth in data has generated the need for new techniques and tools that can intelligently and automatically analyze relevant data. The traditionally method were too time-consuming to cope with the massive quantities of data. As a result, data mining has become a very important research topic, which relies upon powerful computers and efficient algorithms to enhance the value of existing information resources, and can be applied to business management, government administration, scientific and engineering data management and many other areas [30].

Chen et al. [79] provide a comprehensive survey of data mining. They summarized seven requirements and challenges of data mining, which included handling different types of data, efficiency and scalability of data mining algorithms, usefulness, certainty, and expressiveness of data mining results, expression of various kinds of data mining results, interactive mining knowledge at multiple abstraction levels, mining information from different sources of data, and protection of privacy and data security. Some of these requirements may carry conflicting goals. Therefore, different techniques are used to address some of these problems [30].

Data mining is the process of extracting nontrivial relationships from data. No single technique can be defined as the optimal technique for data mining. Fuzzy logic modelling is a probability based method. It has many advantages over the conventional rule induction algorithm, which has been discussed in this chapter. The selection of technique depends on the problem and the data set. How to apply fuzzy logic to data mining problems is still a relatively new field. There are great potentials for exploiting these techniques on different data sets to extract information [30].

In a field of power networks control, a multilayer fuzzy controller has been introduced. An interesting research work has been reported by *Ahmed Rubaai* and *Abdul R. Ofoli* [30].

Transient stability is one of the most important factors studied in power system planning and operation. It is concerned with a system's ability to remain in synchronization following a sudden and major disturbance such as a generator trip or line switching due to faults, or abrupt changes in load generation powers. Several control methods have been devised to handle this problem with some conventional methods being load shedding [80], dynamic braking [81], thyristor controlled series compensator [82], and steam valving and excitation *via* Hamiltonian function method [83, 84]. The conventional approach often requires a precise mathematical model of the controlled systems. In power system practice, there exist parameter uncertainty problems in the plant modelling and, also, since it is large, complex, geographically widely distributed, and can be affected by unexpected events, the conventional controllers often perform satisfactorily over a rather limited range of operation. These reasons also make global control very difficult.

To create flexibility and a strategy that takes into account uncertainties, a solely fuzzy logic control scheme is desirable. Fuzzy control is becoming a powerful control tool in the power industry with application to series-connected flexible alternating current transmission systems (FACTS) devices, wind–diesel power systems, photovoltaic power systems, and thyristor control to the power system stability problem [30].

It was argued that fuzzy control and other allied techniques such as self organizing fuzzy control provide an alternate paradigm to analytic control theory [84].

During the past several years, hierarchical approaches have emerged as one of the most attractive area for research in the application of fuzzy set theory [85 – 88].

Several hierarchical approaches have been directed to design a series of hierarchical fuzzy processors with a small number of input variables distributed in each processor [88]. Other works were based on the development of a hierarchical system of linguistic rules learning methodology [85, 86]. This methodology has been thought of as a refinement of simple linguistic models which, preserving their descriptive power, introduces small changes to increase their accuracy. In this fashion, the knowledge base structure of linguistic fuzzy rule based systems has been extended by introducing the concept of layers.

Aldo Z. Cipriano reported also interesting researches in a field of fuzzy predictive control for power plants [30].

The power generation industry has experienced significant changes in recent decades [89]. Deregulation has brought about a much greater emphasis on economic and financial factors, while growing concerns about the environment have led to restrictions on gaseous emissions and a more intensive search for renewable energy sources. Technological progress has also given rise to major developments in the way of cheaper and more efficient solutions. Generating plants now also have advanced instrumentation and complex computerized monitoring, supervisory and control systems. In the wake of these technological advances, new and greater demands on the control systems have emerged, stimulating the development of new control strategies and algorithms.

In terms of process control, a power plant may be considered as an interacting multi-input multi-output process that is non-linear and time-varying, with disturbances and significant changes in its set-points due to variations in demand. With such characteristics it is impossible to ensure a satisfactory dynamic performance using only PID controllers, and numerous supervisory control strategies have been designed as a result. These strategies maintain PID controllers in the final loops but incorporate more advanced techniques at the higher levels, such as multivariable control, self-tuning adaptive control, robust control, predictive control, fuzzy control, neural networks and genetic algorithms [90].

Fuzzy control also found its application on manufacturing welding systems, in control of weld line of plastic injection-molding [30].

4. Fuzzy Logic in Color Matching and Textile Dyeing Technologies

In this chapter some of the most interesting examples of fuzzy logic usage in colour matching processes will be presented, also an interesting case of fuzzy logic usage in textile dyeing process will be reviewed.

The interesting research performed in the field of textile technology is one by *B. Smith* and *J. Lu* [12]. In their researches they found that non parametric methods of control in textile dyeing process, which includes neural network approach or fuzzy logic approach based on expert systems, are highly appropriate, due to the complexity and uncertainty of dyeing process which is very difficult to control because of the numerous parameters influencing the final result.

Process control is one of the most critical aspects of quality assurance. In textile operations, such as batch dyeing, there are many variables which are under the dyer's control, but many more which are not. Typically these controllable and uncontrollable factors interact in a very complex way.

Traditional manual control methods in textile processes have been automated using microprocessor systems, with corresponding improvement in process repeatability. However, this mode of control utilizes only a minuscule fraction of the total capabilities of modern microprocessor hardware.

In a past few decades, the attempt to improve microprocessor utilization has been realized in wide acceptance of a method of macro scale global linking of information, known as Computer Integrated Manufacturing (CIM) [91, 92]. In typical CIM implementations, sophisticated user interfaces, attractive graphics and computer networking capabilities are employed to make information from machines and machine groups available to managers for real time and post process analysis. The focus is on the use of state of the computing hardware and sensors to acquire data from processes set up common database formats, link islands of automation, and provide management information in a timely manner in a macro or global sense. In CIM, individual workstations are connected through networking paths, and data are stored in a standardized central database, accessible to all workstations. One of the most important features of CIM is that it provides data to managers for strategic decision making. This implementation, although useful in certain ways, is incomplete because it fails to focus on optimized hardware sensor utilization at the micro level, and also because it does not address the under utilization of the microprocessor and other hardware capabilities per se.

So, several innovative concepts have been embodied in a feasibility study presented by Smith and Lu [12]. They suggested possible innovative approaches to dyeing process control scheme, divided into two categories: parametric methods and non – parametric methods. The non – parametric methods include artificial neural network (ANN) control, fuzzy logic (FL) control and expert system (ES) control which is also called knowledge based control or intelligent control.

As the authors pointed out, the traditional dye process control methods attempt to confirm as closely as possible to a specific predetermined process profile (time, temperature, etc.) to achieve correct result. Uncontrollable variances are accepted and, in some cases, reminded after the fact, for example by shade sorting or dye adds.

The authors approach, on the contrary, was to control the ultimate product property of interest (dye shade) by adjusting controllable process parameters in such way to arrive at the desired result. In this approach the process may or may not be the same each time it is run, but the goal is only to arrive at the correct result. The process may be varied each time it is run to compensate for non – controllable factors, such as variances in water quality, substrate preparations, and raw materials.

The controllable parameters that define any dyeing process are temperature, dye concentration, pH, salts concentration. So the authors started from the point of view that the dyeing process is a low order and well damped system, and knowing these characteristics and the control objective, the novel adaptive control algorithm can be designed.

The structure of fuzzy controller proposed by the Smith and Lu can be explained by the simple schematic figure (Figure 8).

It comprises three parts: fuzzifier, rule base, and defuzzifier. A computation of the control action consists of the following stages:

1) Compute current error (E) which the difference between the ideal output and the measured output, as well as its rate of change (CE).
2) Convert numerical E and CE into fuzzy E and CE.
3) Evaluate the control rules using the fuzzy logic operations.
4) Compute the deterministic input required to control the process.

The fuzzifier converts numerical E and CE, such as 2.01, - 0.93, into fuzzy E and CE, such as LN (large negative), MN (medium negative), SN (small negative), ZE (zero), SP (small positive), MP (medium positive) and LP (large positive), with grades of membership m(E) and m(CE) from 0 to 1. The grade of membership values are assigned subjectively to define the meaning of the fuzzy values, such as large negative. The rule base contains the control rules which are developed heuristically for the particular control task and implemented as a set of fuzzy conditional statements of the form: If E is LN and *CE is ZE, then CTRL is MP.* This expression defines a fuzzy relationship between error (E) and error rate (CE) and change of process control input (CTRL) for fuzzy sets.

The defuzzifier is the inverse of the fuzzifier. It converts fuzzy process control input obtained through rule evaluation into numerical deterministic process control input. Many algorithms can be used here, the centre of gravity method being the most popular one.

Simulation results obtained showed that dyeing processes can be controlled effectively using heuristic rules based on fuzzy statements. The controller designer needs detailed knowledge of the dyeing processes in formulating the rules. To get better control performance, other process information like time delay and response speed is also needed. In authors' most recent work, they have developed some enhanced FL schemes which can develop control rules automatically with very limited process knowledge, i.e. the controller has total learning ability. These schemes are now under further evaluation.

Other interesting research regarding colour matching is one performed by *W. Cheetham* and *J. Graf* [13]. In every colour matching process, a number of parameters including also the colorant selection influence the final colour match with customers colour standard. The authors in their research developed the case based reasoning system, the technique that involved fuzzy logic and they showed how different requirements of the colour match are evaluated in order to select the optimum colour formula.

The prior art in colour matching for plastics, but it could be generally applied for every process of colour matching, considered selecting the colorants and loading levels for colour formula, was either accomplished by using working experience or computationally using expensive computer programs. These commercially available computer programs used in colour matching processes use Kubelka – Munk theory in order to convert a set of colorants and loadings into a single colour [7, 24]. The theory describes how the absorption and scattering of colorants in a material are related to the visible colour of the material. Each colorant contributes to the absorption and scattering of the material and its contribution is proportional to the amount of it present in the system multiplied by an absorption and scattering coefficient for that colorant. So, the system proposed by Cheetam and Graf, also uses Kubelka-Munk theory. The system that they proposed was called Form Tool, and has been used from the General Electric Company since the year of its introduction, 1995 [13].

The colour matching basically means the matching of reflectance spectral curves of two or more compared samples, obtained by the spectrophotometric measurement. So the computers used in colour matching processes that contain case – based reasoning software called Form Tool, contains a data of a reflectance curve and a list of pigments and loadings used to create that specific colour. So, how does the colour matching program basically works? It starts from the physical standard that completely satisfy the consumer requirements, which is measured spectrophotometrcally to reads the spectrum and to save it into a colour matching system. If the colour matcher has to provide these specific colours on another sample, he or she will enter the key information such as the resin and grade of material in which to generate the match. Form Tool proposed by the Cheetam and Graf works in a way to search its case – base of previous matches for the "best" previous match and adjust those previous matches to produce a match of the new standards. Of course, there are multiple criteria that the colour match must satisfy. If the acceptable match cannot be found, the Form Tool will adapt this previous match so that it more closely matches the requested colour and application [13].

To achieve the best quality of a final result which is colour match the colour formula must be chosen carefully and the quality of the colour formula must be tested before the production process. Here the Cheetam and Graf used fuzzy logic approach to create the optimal way of finding the formula that will reproduce a specified colour and meet all desired attributes for the application. It means that the degree of a match in all of the attributes desired must be evaluated. Such evaluation needs to provide a consistent meaning of an attributes similarity through all attributes [13]. The authors achieved this consistency through the use of fuzzy linguistic terms, such as excellent, good, fair and poor, which were associated with measured differences in an attribute. Any number of linguistic terms can be used. A fuzzy preference function is used to calculate the similarity of a single attribute of the subject. A fuzzy preference function was used to transform a quantifiable value for each attribute into a qualitative description of the attribute that can be compared with the qualitative description of other attributes. A fuzzy preference function allows a comparison of properties that are based on entirely different scale such as a cost measured in cents per pound and spectral curve match measured in reflectance units [13].

The authors created a various sets of linguistic terms based on discussion with experts. Fuzzy preference function that the authors have been created was defined for each of the following attributes of the colour match: *colour similarity*, *total colour load*, *cost of the colorant formulae*, *optical density of colour* and *colour shift* when molded under normal and abusive conditions [13].

The *colour similarity* means match of two spectral curves compared, and for the best way of colour matching, the colour similarity must be evaluated for all possible lightning conditions. So, comparing spectral curves of objects is the best way to compare their colour, because if the two objects have the same spectral curve, than their colours will match under all lightning conditions. The spectral curve match is characterized by the sums of the squared differences in the reflections values at 31 wavelengths from 400 to 700 nm at a 10 nm interval. The authors created a fuzzy set of linguistic terms excellent, good, fair and poor, and the values of sum of squares that is needed to define these linguistic terms. They also defined the rating of metamerism, which take place when two objects are the same colour under one lightning condition, but different under another lightning conditions. The metamerism index is measured in dE* units using the International Commission on Illuminations L* a* b* colour scale. The standard illuminations that are usually selected when the metamerism index is evaluated are overcast – sky daylight, normal daylight and fluorescent light. The metamerism index is the sum of the three dE* using each of the three illuminants. So the authors also defined the values of metamerism index needed for excellent, good, fair or poor match [7, 13, 24].

The *total colour load* is total volume of all colorants used for a set volume of plastic. It is usually best to use the least volume of colorants that makes an acceptable match. In order to use a fuzzy preference functions for these attributes the case – based must be subdivided into portions that have consistent values for the properties. The authors have divided the case – base into eleven classes corresponding to eleven basic colours: white, grey, black, red, orange, yellow, yellow – green, green, blue – green, blue, violet. An attribute that uses these subclasses is the total loading of colorant in the formula which is dependent upon the colour to be made. So, here also a colour load for each of the eleven colours have been defined to correspond to fuzzy linguistic terms of excellent, good, fair and poor [13].

The *cost* of the colorant in the formula should be kept to a minimum to maximize the profitability of the manufacturing process. The attribute cost is measured in units of cents per pounds and the fuzzy rating for this attribute was specific for particular colour subclasses [13].

The *optical density* of plastic is the thickness of plastic that is required to stop all light from radiating through plastic. The qualitative values of optical density are colour dependent. Fuzzy rating for optical density was also defined for each of the eleven basic colours [13].

The *colour shift* when molding under normal and abusive conditions comes from the fact that the plastic can be molded at low and high temperatures, and than the colour of the plastic change. In order to minimize the colour shift, extra colorant loadings need to be used. The colour of the material is measured under normal processing conditions and under abusive processing conditions, and the differences is measured in dE* units using CIEL*a*b* colour scale. So for each colour subclasses the dE* was defined to correspond to fuzzy linguistic terms of excellent, good, fair and poor [13].

Each of the above properties is based on different scales units. By mapping each of these properties to a global scale through the use of fuzzy preferences and linguistic terms, it becomes possible to compare one attribute to another [13].

V. Bombardier, *E. Schmitt* and *P. Charpetier* developed a vision system according to a fuzzy based sensor concept with aim to improve the wooden board's colour matching [14]. The study performed by the authors concerned the development of a sorting system for wooden boards according to colorimetric aspect. In a production system, the recognition of the wooden board colour is carried out in real time on the industrial production line. This colour identification is usually done by the vision system. The Bombardier, Schmitt and Charpetier considered a concrete industrial problem to carry out their study. The conception of a vision system adapted to the wooden board classification requires the comprehension of the way a human operator works. The vision system developed by the authors can be defined following the fuzzy sensor concept which allows to simplify the adjustment of the vision system settings by a non – specialist. Several attributes are important in an industrial vision system: flexibility, efficiency, reliability, speed, etc. All these attributes are linked to the definition of the inputs and outputs of the system. In case studied by the authors the process is made with colour images according to the following explicit sequence: the acquisition step, the image processing step, the characteristic vector extraction and the decision step (colour grade of a wooden board) [14].

The measurement of the wooden board colour properties are based on colour images. These images are 24 bit RGB images where each component takes value between 0 and 255. To obtain a vision system nearest to human perception, the choice of another colorimetric reference space was necessary. So instead of RGB space the CIEL*a*b* space was chosen because it accurately describes the colour as seen by a human being. After the choice of the representation colorimetric space, a simple characteristic vector has been defined from the industrial constrains (real – time aspect of the data process) [14]. After the choice of the characteristic vector, an adapted classification method was needed. The colours which must be identified are intrinsically fuzzy (impact of the wooden fibre in the wooden colour notion). Thus the descriptors calculated on the images are uncertain but precise. The fuzzy logic allows working with this uncertainty. Moreover, the user of the system expresses his needs under the guise of linguistic terms. Thus the output classes are subjective and non – disjointed. The perception of colour is gradual. For example there are no strict bounds between a “red” wood and a “light red” wood. In this case the difference is given by the wood expert and not by the vision expert. Even if the contrast is not good at the picture, the real hue of the wood board is not evident for a non – specialist. In this way the fuzzy logic concept stands out as the most flexible technique [14].

The colour perception is a very subjective notion and it is strongly linked to the wood species or to its final use. Moreover, different problems appear in an industrial context, for example imprecision of the sensor measurement, acquisition conditions, uncertainty in the colour class definition. In order to take all this into account, the authors decided to develop a colorimetric fuzzy sensor. It is composed of a two parts. The first one aims to deliver a usable measure of the colour from the raw image. Thus, the characteristic vector is obtained after different corrections done in the configuration parts of the sensor. Then these features are injected into a second part of the fuzzy sensor which aims to give the final classification. This decision part is done by a fuzzy rule classifier. As the authors showed, the concept of fuzzy intelligent sensor has several advantages. The fuzzy logic notion allows taking into account the intrinsic wood colour graduality, the imprecision due to the acquisition step and the subjective definition of output colour classes defined by the users. Its generalization capability allows working with small training data sets. The use of a numerical/symbolic converter concept improves the interpretability of the system and finally, the merging of several sensor outputs is made easier and gives better results than other methods. The production system time constraint is also well respected. The main evolution of the proposed system concerns the expansion to other notions than colour. The wooden board appearance is not due to the only concept of hue. Texture and wood grain are also to be taken into account. That is why the authors initiated the development of an appearance fuzzy sensor putting together several wood appearance attributes [14].

S. Gorji Kandi and *M. Amani Tehran* reports research from the field of graphic technologies [15]. They introduce a novel initialization method for fuzzy c-mean algorithm for clustering colour images in graphic technologies.

The colour images contain more information than grey scale ones. In a colour space each colour point is expressed by a three dimensional vector. True colour images typically use one byte for each of the red, green and blue components. So the possible colours of each pixel may contain up to 16.8 millions colours. Due to the limitations of the number of colours of colour image printing and display devices, in many applications, it is preferable to have images with fewer colours as possible and reducing the number of colours is essential. Therefore, colour reduction of images that is usually called colour image quantization is an important process that many researches concentrate on it. Colour image quantization algorithm consists of two major steps: the first step named colour map or colour palette design in which a reduced number of the best possible set of representative colours are chosen; the second step, called pixel mapping in which each colour in the original image is mapped to the best appropriate colour of the colour palette. So the major goal of the colour quantization is to design the colour palette that makes the least perceived difference between original image and quantized one. The algorithms exists for this purpose might be typically categorized in two groups: splitting algorithms that divide the colour space into disjoint regions and clustering based algorithms which perform clustering the colour space into clusters [15].

One of the most popular examples of the clustering based method is the fuzzy c – mean (FCM) algorithm. So in their work the authors proposed a novel initialization method for FCM in the context of colour image clustering. The new proposed method considered usage of principal component analysis (PCA) which was used to find the dominant axis along the colour points to estimate the initial cluster centres in one – dimensional space [15].

As authors have pointed out, the PCA tool can represent the data in a reduced dimensionality space while preserving as much information as possible. The comparison of the proposed method with the typical random initialization method showed that in addition to stability of the results, the quantized images presented by the proposed initialization method are more accurate in terms of S – CIELAB criteria. The S – CIELAB represents colour image difference based upon the CIELAB colour space. There were also some problems about preserving some dominant colour points with low frequency in the original image that can be improved with the proposed method. These colours, which are neglected in colour clustering by FCM algorithm with random initialization method, can be preserved using the proposed initialization [15].

4.1. The Fuzzy Based Approach in Studying the Influence of a Different Textile Surfaces on Colour Similarity

As possibility of using fuzzy logic approach in coloured textile analyses, in this chapter the authors will present the part of the original scientific research on surface structure influences on coloured textiles applying fuzzy logic [96, 97].

As it was said earlier in a report of interesting study of B. Smith and J. Lu [12], the process control is one of the most critical aspects of quality assurance in coloured textile final appearance. The problem in achieving the satisfactory colour matchng and similarity on a different texttiles is that in textile operations, such as batch dyeing, there are many variables which are under the dyer's control, but many more which are not. Typically these controllable and uncontrollable factors interact in a very complex way.

As textile is a highly heterogeneous in its structure, it is often a difficult task to achieve the satisfactory colour similarity on the different structured textile materials. In literature the number of reports about modelling the colour appearance of structured materials can be found [24, 93, 94, 95]. One of the most unconventional approaches was made by Allen and Goldfinger who considered that the best method of modelling it would be through geometric model which considers the fibre geometric characteristics and allows the study of the influence of a change in the form of the coloured material on the reflectance properties [5].

The modern approach to improved colour matching is through application of artificial neural networks (ANN), fuzzy logic control (FL) and experts system control (ES) [12]. Through its application the ANN confirmed its applicability [98, 99] in colour matching in textile dyeings but its require a massively parallel architecture for information processing, which has results in a new paradigm for learning, similar to that in the nervous system. The primary advantage of an ANN controller is its learning ability, which can continuously improve its control performance. The FL and ES control are both used to simulate the decision-making activ ties of an experienced expert. The dif ference between them is the logic. Clas sical expert system control uses crisp logic. Currently, the two approaches are often used together in many cases [12].

In this research performed by authors, the application of fuzzy logic system was examined in aim of achieving the advanced way of observing the influences of surface structure to colour similarity. Fuzzy logic based approach was applied in studying the influence of a different textile surface structure on colour appearance and similarity. It is described how the fuzzy logic based system applies to colour matching and how different requirements of the colour match are evaluated in order to control the optimum colour loading to match the reference sample. The fuzzy based approach has been chosen due to the fact that fuzzy logic provides the computerizing of human reasoning and it provides the ability to handle control problems when there is uncertainty due to complex dynamics of an environment. The advantage of fuzzy logic usage lies in basic characteristics which defines the simplicity and applicability of fuzzy logic system. Fuzzy logic is easy to understand and mathematical concepts behind fuzzy reasoning are simple. It is tolerant of imprecise data and any set of input-output data can be created as a fuzzy system. Fuzzy logic is a method that interprets the values in the input vector and, based on some set of rules, assigns values to the output vector. It is particularly suited to complex system where accurate mathematical models cannot give the satisfactory performance [7, 8, 9, 12, 13, 24]. In this chapter the results of an extensive study will be presented on a one, most indicative, case.

4.1.1. Approach

For the purpose of this research, all measurements were carried out spectrophotometricaly on remission spectrophotometer DataColor® type SF600+CT, using d/8° geometry, with constant instrument aperture (apertures "L"=2.6 cm), with specular component excluded, under illuminant D65, 10 degree observer.

The two 100% cotton knitted samples of the same origin of cotton is chosen for the analysis. The only differences among the samples were in their constructional characteristics which resulted in different surface structures. In the first step of the experimental work a constructional characterization of samples, aiming at précising the importance of the substrate structure and its appearance characteristics influence in colour similarity, was performed. Characteristics provided are shown in Table 3.

Table 3. Characteristics of chosen textile samples

Horizontal density D_h	11.60 stitches/cm
Vertical density D_v	14.00 stitches/cm
Thickness	0.956 mm
Mass per m^2	283 g/m^2
Yarn fineness	basally fibre – 24 tex (front side) coating fibre – 59 tex (back side)

Also the surfaces of the chosen samples were scanned with the CCD digital camera. The scans of samples specific surface structure are shown on Figure 9.

Sample 1

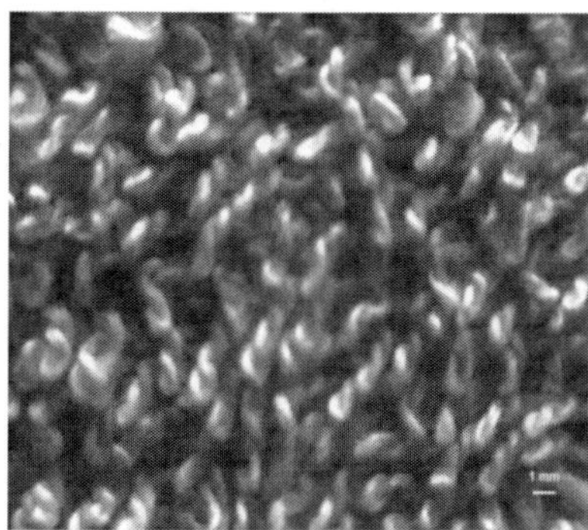

Sample 2

Figure 9. CCD Digital camera scans of the samples.

First the both samples were dyed in four basic concentrations of dyestuff: 0.25%, 0.75%, 1.4% and 2%. Afterward, samples were measured spectrophotometrically and the K/S values were compared in order to establish is there any difference in equally dyed samples regarding the specific surface structure (Figure 10).

Further, the sample 1 was chosen as the standard (reference sample). Its data, obtained by the spectrophotometer measurement, were stored in a computer database and were used as the reference data in a further work. The computer matching (CMP) on sample 2 according to reference sample was performed, for determining the suitable concentration of dyestuff in relation to specific surface spectral characteristics of sample 2. A yellow shaded direct dyestuff suitable for cotton dyeing CI Direct Yellow 50 was used in experimental work. For usage of this technique the computer database of spectrophotometer data for chosen dyestuff and for each type of substrate used in the experimental work are required.

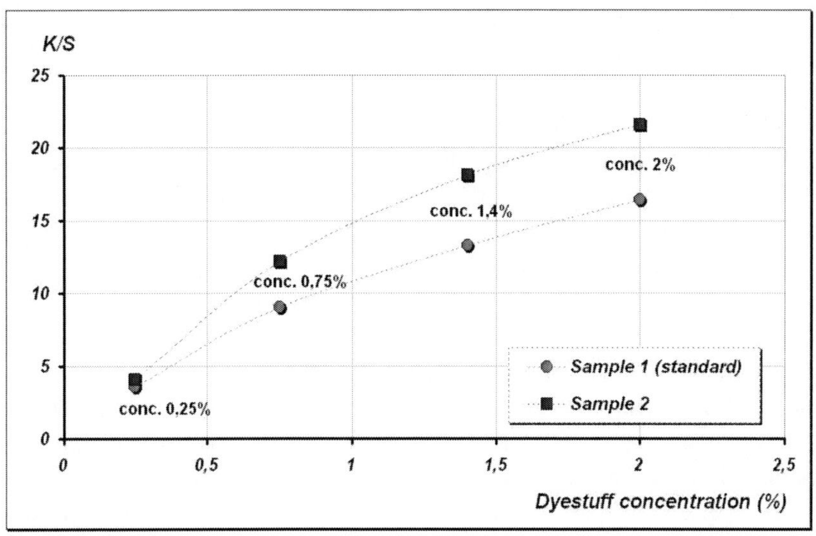

Figure 10. K/S values obtained for samples 1 and 2 regarding the dyestuff concentration used.

In the next step of the experimental work the fuzzy logic system was created with the input space based on terms that would define the surface characteristics of samples and through the set of rules, provide the output space which would be the change of dyestuff concentration in dependence of the characteristics defined for one chosen sample. The fuzzy logic system was constructed using MATLAB and Fuzzy Logic Toolbox.

The input space was created based on five linguistic variables – defined as thickness, roughness, gloss, thread density and the dyestuff concentration. For the input linguistic variables – thickness, roughness, gloss and thread density – the 3 linguistics values were used: low, middle and high. For the input linguistic variable – dyestuff concentration – the 6 linguistics values were used: very low, low, middle, middle high, high and very high. For the output space definition, the one linguistic variable was chosen based on the percentage of basic dyestuff concentration change in dependence of input linguistic variables values assign for one tested sample. The simple scheme of the input and the output space of the fuzzy logic reasoning constructed, is shown on Figure 11.

For each linguistic variable the membership functions and their parameters were defined for each linguistic value according to the experimental knowledge. The degree to which the inputs belong to each of the appropriate fuzzy sets is determined via membership functions. Three built-in membership functions were used (smf, zmf and pimf). Smf, zmf and pimf are names of built-in membership functions of MATLAB Fuzzy Logic Toolbox. First one is smf function. Its syntax is y=smf (x,[a b]). This spline based curve is a mapping on the vector x, and is named because of its S-shape. The parameters *a* and *b* locate the extremes of the sloped portion of the curve. The second one is zmf function. Its syntax is y=zmf (x,[a b]). This spline based function of x is so named because of its Z-shape. The parameters *a* and *b* locate the extremes of the sloped portion of the curve. The third one is pimf function. Its syntax is y=pimf (x,[a b c d]). This spline-based curve is so named because of its π-shape. This membership function is evaluated at the points determined by the vector x. The parameters *a* and *d* locate the "feet" of the curve, while *b* and *c* locate its "shoulders". It is a combination (multiplication) of two above mentioned functions: smf and zmf. Its syntax, because of that, is also y=smf (x,[a b]).*zmf (x,[c d]) [5]. Their graphs are given in Figure 12.

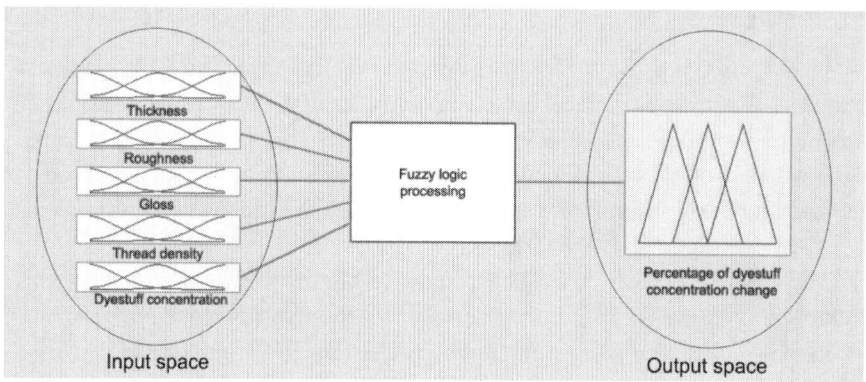

Figure 11. Schematic description of a fuzzy logic system.

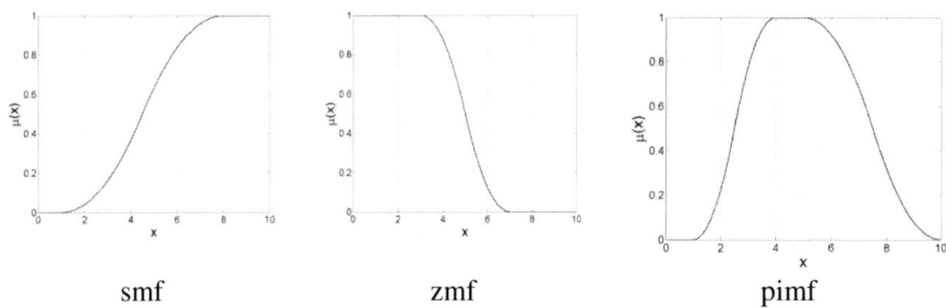

Figure 12. *smf* membership function with parameters a=1 and b=8; *zmf* membership function with parameters a=3 and b=7; *pimf* membership function with parameters a=1, b=4, c=5 and d=9 is shown.

The values of *a* and *b* or values of *a*, *b*, *c* and *d* parameters of each function is defined based on expertise and experimental knowledge which lies on a number of experimental dyeings performed in the range of dyestuff concentration from 0.05% to 2%, used for dyestuff concentration estimation and fuzzy logic system adjustment.

The values of *a* and *b* or values of *a*, *b*, *c* and *d* parameters for each membership function are shown in Table 4 and 5.

Sixty (60) if–then rules were created following the example: If *thickness* is *low* and *roughness* is *middle* and *gloss* is *low* and *thread density* is *low* and *dyestuff concentration* is *very low* then *change of basic dyestuff concentration* is *middle*. For the implication the two built-in methods from Fuzzy Logic Toolbox are supported: min (minimum) and prod (product) [5]. In this paper the min methods for implication is used. The min method for AND operator and max method for OR operator were used and also the max method for aggregation was used. For deffuzification the centroid method was used. These 60 if–then rules were also defined based on experimental knowledge.

Finally the performance of both systems were compared and analyzed. In this work the approach are based exactly on surface – structural characteristics of a textile samples and the aim was to analyze the applicability of fuzzy logic approach which would provide the satisfactory incorporation of these characteristics in colour matching technique. The main idea was to examine the approach using fuzzy logic based technique in order to provide the method that would include surface parameters of textile samples that are of importance in colour applying on a different structured textile surfaces.

Table 4. Parameters *a*, *b*, *c* and *d* for membership function of each input linguistic variable

Input linguistic variable – thickness		Range 0.3-1.1
Linguistic value	Function	Parameters
low	zmf	a=0.41; b=0.60
middle	pimf	a=0.53; b=0.62; c=0.75; d=0.87
high	smf	a=0.84; b=0.94
Input linguistic variable – roughness		Range 0.0-3.0
Linguistic value	Function	Parameters
low	zmf	a=0.43; b=1.21
middle	pimf	a=0.52; b=1.24; c=1.88; d=2.49
high	smf	a=1.95; b=2.58
Input linguistic variable – gloss		Range 0.0-3.0
Linguistic value	Function	Parameters
low	zmf	a=0.52; b=1.08
middle	pimf	a=0.66; b=1.22; c=1.89; d=2.44
high	smf	a=2.02; b=2.62
Input linguistic variable – threads density		Range 0.0-1.0
Linguistic value	Function	Parameters
low	zmf	a=0.15; b=0.39
middle	pimf	a=0.18; b=0.40; c=0.53; d=0.72
high	smf	a=0.65; b=0.83
Input linguistic variable – range of dyestuff concentration		Range 0.0-5.0
Linguistic value	Function	Parameters
very low	pimf	a=0; b=0.18; c=0.29; d=0.44
low	pimf	a=0.34; b=0.63; c=0.78; d=1.12
middle	pimf	a=0.91; b=1.19; c=1.50; d=1.86
middle high	pimf	a=1.60; b=1.90; c=2.37; d=2.72
high	pimf	a=2.51; b=2.77; c=3.07; d=3.46
very high	smf	a=3.18; b=3.57

Table 5. Parameters *a*, *b*, *c* and *d* for membership function of output linguistic variable

Output linguistic variable – assumed percentage of aimed concentration change		Range 0.0-150.0
Linguistic value	Function	Parameters
extremely low	zmf	a=3.8; b=10.0
very low	pimf	a=5; b=10.9; c=18.1; d=25
middle low	pimf	a=20; b=25.7; c=33.3; d=40
Low	pimf	a=35; b=41.3; c=48.45; d=55
Middle	pimf	a=50; b=56; c=63.6; d=70.3
middle high	pimf	a=65; b=69.5; c=75.54; d=80
High	pimf	a=75; b=81.4; c=87.5; d=95
middle very high	pimf	a=90; b=95.5; c=103; d=115
very high	pimf	a=105; b=112.8; c=120; d=130
extremely very high	smf	a=125; b=131.9

4.1.2. Results and Discussion

For samples composed of the same materials variances may be seen in colour due to differences in a smoothness or gloss of the surface. Colour of textile surfaces may appear different if the surface conditions are changed, because the amount of specular reflectance and diffused reflectance changes. Since the human eye view only the diffused light, it is expected that, for the viewer, the differences in surface quality will influence the difference appearance of colour. So it can be said, that because of certain subjective elements, human eye recognized the changes in surface as the changes in colour. The aim was to find a way to include subjective elements that influence the colour appearance into a studying the colour matching and similarity.

Observing the results of samples dyed in four basic concentrations through K/S value (Fig. 10), it can be seen that the surface of sample 1 (smoother surface) produce lower K/S value of the same amount of dyestuff, as the result of the higher reflectance from the smoother surface. It is in correlation with the K/S value obtained higher for sample 2 (rougher surface) which confirm the influence of surface structure on appearance of colour and showed that the rougher surface, even equally coloured, will perform deeper colour. Based on results, which indicate the necessity of a dyestuff concentration correction (Fig. 10), the "CMP" procedure was performed. For sample 2 the dyestuff concentration correction was performed based on the reference sample data (sample 1) stored in database. The corrected concentrations are shown in Table 6.

Table 6. Dyestuff concentrations provided by CMP system based on corresponding concentration of reference sample

Basic dyestuff concentrations (%)	**0.25**	**0.75**	**1.4**	**2**
New concentrations obtained by the CMP operation (%)				
Sample 2	**0.22%**	**0.71%**	**1.35%**	**1.89%**

It must be clarified that the colour parameter values of the chosen dyestuff used in the experimental work from the input fuzzy logic system space were absent. Those parameters were not included in the input fuzzy logic space because of the complexity of colour definition. The corrections were made based on the change in input variable dyestuff concentration expressed as a percentage (%), which was defined based on the number of experimental dyeings performed in different ranges of dyestuff concentration on different structured cotton material. The fuzzy logic system does not give any answer on colour, but shows, based on a set of rules, how to make the correction, depending on the different surface structures. Therefore, the colour here was used only as control, while applying the same colour on different structured textiles.

The output variable provided the results of concentration change which are shown in Table 7.

Table 7. Dyestuff concentrations provided by fuzzy logic system based on corresponding concentration of computer base reference sample in dependence of characteristics of a specific surface structure of tested samples

Basic dyestuff concentrations (%)	**0.25**	**0.75**	**1.4**	**2**
New concentrations obtained by the fuzzy logic system (%)				
Sample 2	**0.165%**	**0.69%**	**1.21%**	**1.50%**

Comparing the results obtained in Table 6 and 7, it can be seen that discrepancies of corrected concentrations obtained by the CMP system (Table 6) are lower on average in compare to those provided by fuzzy system (Table 7). The reason is that in CMP system the concentrations are provided based strictly on surface-spectral characteristics provided by spectrophotometric measurement. The results provided by CMP system arising from the interaction of parameters that influence the spectral-surface remission characteristics. Due to such characteristics the CMP colour matching system provided slight differences in concentration change, since its dealing only with the surface – spectral phenomena.

In fuzzy logic system all subjective and objective parameters can be processed by proper assigning of linguistic values and correct determination of membership function, which is vital in assigning inputs to appropriate fuzzy set and are necessary for proper system performance. So, due to the different linguistic value assigned for the linguistic variable – roughness in fuzzy logic system, the distinguished difference in percentage of dyestuff concentration change occurred.

To compare the CMP system to constructed Fuzzy Logic System (FLS) system the samples were dyed according to concentrations provided by both system. After the dyeing, dried samples were measured spectro-photometrically. The results obtained in both dyeings were compared to the reference sample 1 and the total colour difference values (dE_{CMC}) were calculated using the CMC (l:c) colour formula. The results are presented in a form of a bar chart given on Figure 13.

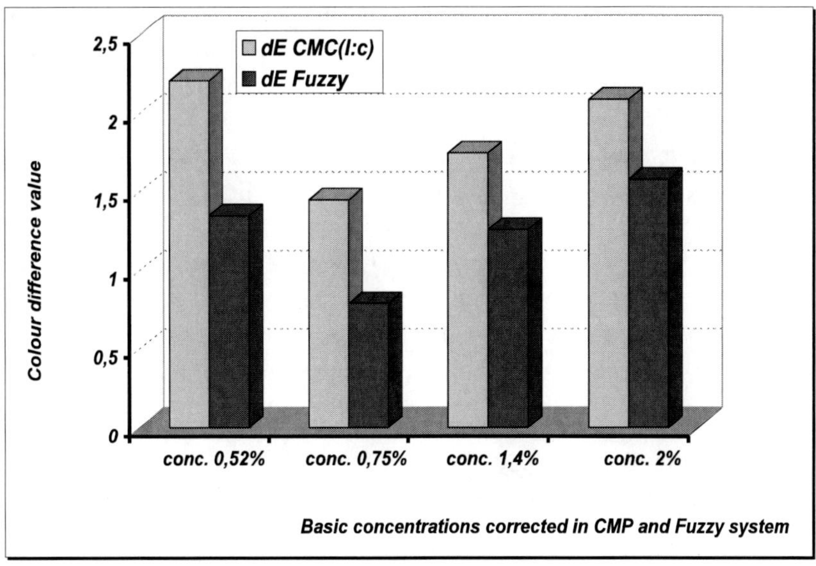

Figure 13. Difference values obtained for sample 2 compared to reference sample 1, for four basic dyestuff concentrations corrected in CMP and Fuzzy system.

The colour differences (dE_{CMC}) obtained for fuzzy logic system are lower and closer to value of 1 (in the range of agreed tolerances which is for textile set on $dE_{CMC} \leq 1$) comparing to those obtained with CMP system. The results obtained confirms the possibility of using the linguistic form of fuzzy logic that allows any subjective as well as objective parameter to be processed, which is of significant importance in studying the influence of textile surface structure parameters on colour appearance.

4.1.3. Conclusion

The possibility of fuzzy logic based reasoning application is confirmed, but further study of the subject is necessary with detailed analyze of a fuzzy logic rule construction.

The CMP approach concerns surface phenomena based on spectrophotometric measurement of surface–spectral remission and it does not concern other parameters that would influence the visual experience of an observer. On the other hand, the results obtained for Fuzzy system confirmed that the fuzzy based system can react on change of surface structure parameters, which is result of proper definition of rules and defined linguistic variables that use words for assigning values of properties that influence the experience of an observer (gloss, smoothness, roughness, etc.).

So, the results obtained confirm that the fuzzy logic reasoning certainly have its application in control of surface structure influence of coloured textiles, especially for lighter shades where the influence of structure parameters on the colour experience of an observer is more emphasized.

Certain limitations of this procedure do exists and are that concrete fuzzy system is applicable only for this particular colour - yellow, and for each other colour, separate fuzzy system should be constructed.

Also it requires a lot of experimental knowledge and experience. The analysis and the results shown in this paper are the part of a research but they indicate the further research since it has been confirmed that fuzzy based system can be applied for surface structure influence evaluation and control.

REFERENCES

[1] Tong, R. M.: An Annotated Bibliography of Fuzzy Control, *Industrial Applications of Fuzzy Control*, M. Sugeno, Ed., North- Holland, Amsterdam, Holland, 1985, pp. 249-269.

[2] Terano, T.; Asai, K.; Sugeno, M.: *Fuzzy Systems Theory and Its Applications*. Academic Press, Boston, MA, USA.

[3] Pedrycz, W.: *Fuzzy Control and Fuzzy Systems*, 2nd extended ed. Wiley, New York, NY, USA.

[4] Berenji, H.: Fuzzy Logic Controllers, *An Introduction to Fuzzy Logic Applications in Intelligent Systems*, R. R. Yager and L. A. Zadeh, Eds., Kluwer Academic Publishers, Boston, MA, USA, 1992, pp. 69-96.

[5] Zadeh, L. A.: Fuzzy Logic Toolbox for Use with Matlab, The Math Works Inc., Natick, Ma, USA, 1995, pp. 4 – 28.

[6] Lu, j; Smith, B.: Improving Computer Control of Batch Dyeing Operations, *American Dyestuff Reporter*, No. 9, (1993), pp. 17 – 35.

[7] Grundler, D.; Rolich, T.; Glogar, M. I.: Imrpoving Textile Colour Matching Using Fuzzy Logic, *Book of Papers,* MIPRO 2007 International Conference, Opatija, Croatia, (2007), pp. 135 – 138.

[8] Wenstop, F.: Quantitative Analysis with Linguistic Values, *Fuzzy Sets and System*, No. 4., (1980), pp. 99 – 115.

[9] Dubois, D.; Prade, H.: Fuzzy Sets and System: Theory and Application, Academic Press, New York, (1980).

[10] Westland, S.; Ripamonti, C.: Computational Colour Science Using Matlab, John Wiley & Sons Ltd, The Atrium, Southern Gate, Chichester, West Sussex, England, ISBN 0-470-84562-7 (2004).

[11] Dubois, D.; Prade, H.: Fundamentals of Fuzzy Sets, Kluwer Academic Publisher, Norwell Massachusetts, USA, (2000), ISBN 0-7923-7732-X.

[12] Smith, B.; Lu, J.: Improving Computer Control of Batch Dyeing Operations, American Dyestuff reporter, September, (1993.), pp 17 - 36.

[13] Cheetam, W; Graf, J.: Case – Based Reasoning in Color Matching, Proceedings of the Second International Conference on Case – Based Reasoning Research and Development, (1997), ISBN 3-540-63233-6, pp. 1 – 12.

[14] Bombardier, V.; Schmitt, E.; Charpentier, P.: A Fuzzy Sensor for Color Matching Vision System, Measurement 42 (2009), pp. 189 – 201.

[15] Kandi Gorji, S.; Tehran Amani, A.: Obtaining Proper Initial Cluster for the Fuzzy C – Mean Algorithm for Color Reduction, Proceedings of Midterm Meeting of the International Color Association, Hangzou, China, (2007), pp. 110 – 113. .

[16] Tsoutseos, A. A.; Nobbs, J. H.: Alternative Approach to Colour Appearance of Textile Materials with Application to the Wet/Dry Reflectance Prediction, *Textile Chemists and Colourists & American*

Dyestuff Reporter, Vol. 32, No. 6, ISSN 0040-490X, (2000), pp. 38 – 43.

[17] Gerla, G.: Fuzzy Logic: Mathematical Tools for Approximate Reasoning, Kluwer Academic Publisher, Dordrecht, (2001), ISBN 0-7923-6941-6.

[18] Zadeh, L. A.: Outline of a New Approach to the Analyses of Complex Systems and Decision Processes, *IEEE Transaction on Systems, Man and Cybernetics*, 3(1)1973, pp. 28 – 44.

[19] Zadeh, L. A.: Fuzzy Algorithms, *Information and Control*, 12(1968), pp. 94 – 102.

[20] Zadeh, L. A.: Making Computers Think Like People, *IEEE Spectrum*, 8(1984), pp. 26 – 32.

[21] Korner, S.: Laws of Thought, Ecyclopedia of Philosophy, Vol. 4, MacMillan, NY, (1967), pp. 414 – 417.

[22] Lejewski, C.: Jan Lukasiewicz, Encyclopedia of Philosophy, Vol. 5, MacMillan, NY, (1967), pp. 104 – 107.

[23] Hellman, M.: Fuzzy Logic Introduction, Laboratorie Antennes Radar Telecom, Universit'e de Rennes, France; http://epsilon.nought.de/ tutorials/fuzzy/fuzzy.pdf.

[24] Glogar, M. I.: Study of Optical Properties of Coloured Textile Surfaces and "CMP" Operation Application, Ph.D. Thesis, University of Zagreb Faculty of Textile Technology, (2006).

[25] Fuzzy Logic Fundamentals, http://www.ece.msstate.edu/courses/ ece4723/fall11/notes/ FL_chapter.pdf.

[26] Jantzen, J,: Tutorial on Fuzzy Logic, Technical University of Denmark, Department of Automation, BLDG 326, DK – 2800 Lyngby, Denmark, (1998).

[27] Yager, R.: General Class of Fuzzy Connectives, *Fuzzy Sets and System,* 4(1980), pp. 235 – 242.

[28] Schweizer, B.; Sklar, A.: Associative Functions and Abstract Semi – Groups, Publ. Math Debrecen, 10 (1963), pp. 69 – 81.

[29] Sugeno, M.: Fuzzy Measures and Fuzzy Integrals: A Survey, (M. M. Gupta, G. N. Saridis and B. R. Gaines, editors), *Fuzzy Automata and Decision Processes*, North Holland, New York, (1977), pp. 88 – 102.

[30] Bai, Y.; Zhuang, H.; Wang, D.: Advanced Fuzzy Logic Technologies in Industrial Applications, Springer – Verlag London Limitted, London UK, (2006), ISBN 1-84628-468-6.

[31] Mamdami, E. H.; Assilian, S.: An Experiment in Linguistic Synthesis with a Fuzzy Logic Controller, *International Journal of Man Machine Studies*, 7(1)1975, pp. 1 – 13.

[32] Kickert, W.; Mamdami, E. H.: Analysis of a Fuzzy Logic Controller, *Fuzzy Set and System*, 12(1978), pp. 29 – 44.

[33] Sugeno, M. (Ed.): Industrial Aplication of Fuzzy Logic Control, North Holland, Amsterdam Holland, 1985.

[34] Yasunobu, S.; Miyamoto, S.: Automatic Train Operation by Predictive Fuzzy Control, *Industrial Application of Fuzzy Logic*, M. Sugeno (Ed.), Amsterdam. North Holland, (1985), pp. 1 – 8.

[35] Yager, R.; Filey, D.: Essentials of Fuzzy Modeling and Control, J. Wiley, (1994), ISBN 0471017612.

[36] Osta, A.; De Gloria, A.; Farabochi, P.; Pagni, A.; Rizzotto, G.: Hardware Solutions for Fuzzy Control, Proceedings of the IEEE, 83(3)1995, pp. 422 – 434.

[37] Mendel, M. J.: Fuzzy Logic Systems for Engineering; Proceedings of the IEEE, 83(3)1995, pp. 345 – 377.

[38] Lau, K.; Hocken, R.; Haight, W.: An Automatic Laser Tracking Interferometer System for Robot Metrology, National Bureau of Standards, Gaithersburg, MD, (1985), pp. 113 – 152.

[39] Harb, S., M.: Laser Tracking System for Multi – Axis Machine Calibration, *International Conference on Laser Metrology and Machine Performance*, Southampton, England, (1995), pp. 161 – 171.

[40] Nakamura, O.; et al.: A laser Tracking Robot – Performance Calibration System Using Ball – Seated Bearing Mechanisms and a Spherically Shaped Cat's – Eye Retroreflector, Review of Scientific Instruments, 65(4)1994, pp. 1006 – 1011.

[41] Rene, R. J.: A Portable Instrument for 3D Dynamic Robot Measurements Using Triangulation and Laser Tracking, IEEE Transactions on Robotic and Automation, 10(4)1994, pp. 504 – 516.

[42] Bao, Y.; Fujiwara, N.: Dynamic Measurement Orientation by LTS, Proceedings of the Japan –USA Symposium on Flexible Automation, (1996), pp. 545 – 548.

[43] Vincze, M.; Prenninger, J. P.; Gander, H.: A Laser Tracking System to Measure Position and Orientation of Robot End – Effectors Under Motion, The International Journal of Robotics Research, 13(4)1994, pp. 305 – 314 .

[44] Spiess, S.; Vincze, M.; Ayromlou, M.: On the Calibration of a 6D Laser Tracking System for Contactless Dynamic Robot Measurements, IEEE

Instrumentation and Measurements Technology Conference, Ottawa, Canada, (1997), pp. 1203 – 1208.

[45] Leigh – Lancaster, C. J.: Development of a Laser Tracking System, Proceedings of Fourth Annual Conference on Mechatronics and Machine Vision in Practice, Toowoomba, Australia, (1997), pp. 163 – 168.

[46] Shieh, M. Y.; Li, T. H. S.: Implementation of Integrated Fuzzy Logic Controller for Servomotor System, IEEE International Conference on Fuzzy System, 1995, pp. 1755 – 1762.

[47] Kawaji, S.; Matsunaga, N.: Design of Fuzzy Model Following Servo Control Systems, *IEEE International Cnference on Fuzzy System*, (1995), pp. 1763 – 1768.

[48] Betin, F.; Pinchon, D.; Capolino, G. A.: Fuzy Logic Applied to Speed Control of a Stepping Motor Drive, *IEEE Transaction on Industrial Electronics*, 47(3)2000, pp. 610 – 621.

[49] Ha, P. Q.; Hguyen, H. Q.; Rye, D. C.; Durrant – Whyte, H. F.: Fuzzy Sliding – Mode Controllers with Applications, *IEEE Transaction on Industrial Electronics*, 48(1)2001, pp. 38 – 45.

[50] Li, C.; Lee, C. Y.: Fuzzy Mozion Control of an Auto – Warehousing Crane System, *IEEE Transaction on Industrial Electronics*, 48(5)2001, pp. 983 – 993.

[51] Li, W.; Chang, X. G.; Farrel, J.; Wahl, F. M.: Design of an Enhanced Hybrid Fuzzy P+ID Controller for a Mechanical Manipulator, *IEEE Transactions on System, Man and Cybernetics – Part B: Cybernetics*, 31(6)2001, pp. 938 – 945.

[52] Rubaai, A.; Ricketts, D.; Kankam, M. D.: Laboratory Implementation of a Microprocesor – Based Fuzzy Logic Tracking Controller for Motion Control and Devices, *IEEE Transaction on Industrial Application*, 38(2)2002, pp. 448 – 455.

[53] Ronsefeld, A.: Survey: Image Analysis and Comuter Vision; 1993, *Computer Vision and Image Understanding*, 59(3)1994, pp. 367 – 404.

[54] Suarez, J. I.; Vinagre, B. M.; Chen, Y. Q.: Spatial Path Tracking of an Autonomous Industrial Vehicle Using Fractional Order Controllers, ICAR 2003 Congress, Prtugal, (2003).

[55] Rossetter, E. J.; Gerdes, J. C.: Performance Guarantees for Hazard Based Lateral Vehicle Control, Proceedings of the 2002 IMECE Cnference, (2002).

[56] Bentalaya, S. et al.: Fuzzy Path Tracking Control of a Vehicle, IEEE International Conference on Intelligent Vehicles, (1998), pp. 195 – 200.

[57] Holve, R.; Protzel, P.; Naab, K.: Generating Fuzzy Rules for the Acceleration Control of an Adaptive Cruise Control System, Fuzzy Informaton Processing Society, (1996), NAFIS Bienal Conference of the North American, pp. 451 – 455.

[58] Takagai, T.; Sugeno, S.: Fuzzy Identification of System and Its Application to Modelling and Control, *IEEE Transactions on Systems, Man and Cybernetics*, 15(1)1985, pp. 116 – 132.

[59] Yashimura, H.; Hirako, A.: Automated Mechanical Transmission Control, Proceedings of the International Congress on Transportation Electronics, Deaborn, USA, (1986), pp. 179 – 189.

[60] Oin, G; Fan, J.; Zhang, H.; Ge, A.: Microcomputer Control System of Automated Mechanical Transmission, *Chinese Automotive Engineering*, 21(1)1999, pp. 21 – 25.

[61] Jo, H. S.; Jo, S. T.; Lee, J. M.; Park, Y. I.: Analysis of the Shift Characteristics of a Automated Manual Transmission in the Parallel Type Hybrid Drive Train System of a Transit Bus, *Heavy Vehicle System*, 8(1)2001, pp. 60 – 82.

[62] Fredriksson, J.; Egardt, B.: Nonlinear Control Applied to Gear Shifting in Automated Manual Transmissions, Proceedings of the IEEE Conference on Decission and Control, Vol. 1, Sydney, Australia, (2000), pp. 444 – 449.

[63] Tanaka, H.; Wada, H.: Fuzzy Control of Cluth Engagement for Automated Manual Transmission, *Vehicle System Dynamics*, Vol. 24, (1995), pp. 4 – 8.

[64] Nguyen, D.; Widrow, B.: The Truck Backer – upper: An Example of Self – learning in Neural Network, *IEEE Control System Magazine*, 10(2)1990, pp. 18 – 23.

[65] Wang, L. X.; Mendel, J. M.: Generating Fuzzy Rules by Learning from Examples, *IEEE Transaction on System, Man and Cybernetics*, 22(6)1992, pp. 1414 – 1427.

[66] Ramamoorthy, P. A.; Huang, S: Fuzzy Expert System vs. Neural Networks – Truck Backer – Upper Control Revisited, Proceedings of IEEE International Conference on Robotics and Automation, (1991), pp. 221 – 224.

[67] Shann, J. J.; Fu, H. C.: A Fuzzy Neural Network for Rule Acquiring on Fuzzy Control Systems, *Fuzzy Sets and Systems*, Vo. 71, (1995), pp. 345 – 357.

[68] Kim, D.: Improving the Fuzzy System Performance by Fuzzy System Ensemble, *Fuzzy Sets and System*, Vol. 98, (1998), pp. 43 – 56.

[69] Dumitrache, I.; Catalin, B.: Genetic Learning of Fuzzy Controllers, *Mathematics and Computer in Simuation*, Vol. 49, (1999), pp. 13 – 26.

[70] Saffiotti, A.: The Uses of Fuzzy Logic in Autonomous Robot Navigation, *Journal of Soft Computing*, 1(4)1997, pp. 180 – 197.

[71] Saffiotti, A.: Fuzzy Logic in Autonomous Robotics: Behaviour Coordination, Proceedings of the 6th IEEE International Conference on Fuzzy System, Barcelona, Spain, (1997), p. 573 – 578.

[72] Tunstel, E.; Lippincotti, T.; Jamshidi, M.: Behaviour Hierarchy for Autonomous Mobile Robots: Fuzzy Behaviour Modulation and Evolution, *International Journal of Intelligent Automation and Soft Computing*, 3(1)1997, pp. 37 – 49.

[73] Kadmiry, B.; Bergsten, P.; Drankov, D.: Autonomous Helicopter Control Using Fuzzy Gain Scheduling, Proceedings of the 2001 IEEE International Conference on Robotics & Automation, Seoul, Korea, (2001), pp. 2981 – 2985.

[74] Tanaka, K.; Sugeno, m.: Stability Analysis and Design of Fuzzy Control Systems, *Fuzzy Sets and System*, 45(2)1992, pp. 135 – 156.

[75] Kadmiry, B.; Driankov, D.: Autonomous Helicopter Control Using Linguistic and Model – Based Fuzzy Control, Proceedings of the 2001 IEEE International Symposium on Intelligent Control, Mexico City, (2001), pp. 348 – 352.

[76] Maedea, Y.; Takegami, M.: Collision Avoidance Control Among Moving Obstacles for a Mobile Robot on the Fuzzy Reasoning, *Journal of the Robotic Society of Japan*, 6(6)1988, pp. 518 – 522.

[77] Takeuchi, T.: An Autonomous Fuzzy Mobile Robot, *Journal of the Robotic Society of Japan*, 6(6)1988, pp. 536 – 541.

[78] Murofushi, T.; Sugeno, M.: Fuzzy Control of Model Car, *Journal of the Robotic Society of Japan*, 6(6)1988, pp. 542 – 555.

[79] Chen, M. S.; Yu, P. S.: Data Mining: An Overview from a Database Perspective, *IEEE Transaspect Knowledge and Data Engineering*, Vol. 8, (1996), pp. 866 – 883.

[80] Dicorato, M.; Scala, M. L.; Scarpellini, P.: A Corrective Control for Angle and Voltage Sability Enhancement on the Transient Time – Scale, *IEEE Trans. Power Syst.,* Vol. 15, (2000), pp. 1345 – 1353.

[81] Rubaai, A.; Ofoli, A. R.: Design and Analysis of Nonlinear Digital Controllers – based Two – level Hierarchy for Electric Utility Industry, *IEEE Trans. Ind. Applicant,* Vol. 39, (2003), pp. 395 – 407.

[82] Son, M. K.; Park, K. J.: On the Robust LQG Control of TCSC for Damping Power System Oscillations, *IEEE Trans. Power Syst.,* Vol. 15, (2000), pp. 1306 – 1312.

[83] Xi, Z.; Cheng, D.; Lu, Q.; Mei, S.: Nonlinear Decentralized Controller Design for Multi – machine Power System Using Hamiltonian Function Method, *Automatica*, 38(3)2002, pp. 527 – 534.

[84] Mamdami, H. E.: Twenty Years of Fuzzy Control: Experiences Gained and Lessons Learnt, Proceedings of 2nd IEEE Conference of Fuzzy Systems, Vol. 1, (1993), pp. 339 – 344.

[85] Cordon, O.; Herrera, F.: Linguistic Modelling Hierarchical Systems of Linguistic Riles, *IEEE Trans. Fuzzy Sets*, Vol. 10, (2002), pp. 2 – 20.

[86] Yager, R. R.: On the Construction of Hierarchical Fuzzy Systems Models, *IEEE Trans. Syst., Man., Cybern.,* Vol. 28, (1998), pp. 55 – 66.

[87] Gegov, A. E.; Frank, M. P.: Hierarchical Fuzzy Control of Multivariable Systems, *Fuzzy Sets Syst.,* Vol. 72, (1995), pp. 299 – 310.

[88] Raju, C. V. S.; Zhou, J.: Adaptive Hierarchical Fuzzy Controller, *IEEE Trans. Syst., Man, Cybern.,* Vol. 23, (1993), pp. 973 – 980.

[89] Flynn, D.: Thermal Power Plant Simulation and Control, *IEE Power & Energy Series*, Vol. 43, (2003).

[90] Saez, D.; Cipriano, A.; Ordys, A. W.: Optimisation of Industrial Processes at Supervisory Level: Application to Control of Thermal Power Plants, Springer, London UK, (2002), ISBN 978-1-85233-386-7.

[91] Banky, P. G.: Flexible Manufactirung Cells and Systems, CIMware Limited, Guilford, England UK.

[92] Scheer, A. W.: CIM – Toeards Factory of the Future, 2nd ed. Springer – Verlag, New York USA, (1991).

[93] Mourad, S.; Emmel, P.; Simon, K.: Extending Kubelka - Munk`s Theory with Lateral Light Scattering, Proceedings of International Conference on Digital Printing Technologies, Florida, USA, (2001).

[94] Joaneli, M. I.; Parac – Osterman, Đ; Golob, D.: Textile Surface Structure Importance and Kubelka – Munk Theory in Colour Match Calculations, *Colourage, Vol.53, No. 4., ISSN: 0010 – 1826 – 59 – 67* (2006).

[95] *Tominaga, S: Szrface Reflectance Estimation by the Dichromatic Model, Colour Research Application*, Vol. 21, No. 2, (1996), pp. 104 – 114.

[96] Glogar, M. I.; Parac – Osterman, Đ.; Grundler, D.; Rolich, T.: Research of Surface Structure of Coloured Textiles: Applying Fuzzy Logic, *Coloration Technology*, **127** (2011) 6; 396 – 403, ISBN: 1472 – 3581.

[97] Glogar, M. I.; Parac – Osterman, Đ.; Grundler, D.; Rolich, T.: The Rule-Based Approach in Studying the Influence of a Different Textile Surfaces on Colour Similarity, *Proceedings of the 22nd International DAAAM Symposium "Intelligent Manufacturing & Automation"*, Vienna, Austria, 2011; 1207 – 1208, ISBN: 978-3-901509-83-4.

[98] Kim, J.; McGregor, R.: Theoretical Modelling of Batch Dyeing Process, *Report of CRAFTM Sponsor Meeting*, NCSU, 1991.
Spiekermann, C.; Lu, J.; Smith, C. B.; McGregor, R.: A Novel Approach for Modelling and Controlling Dyeing Processes, *AATCC International Dyeing Symposium*, Charlotte, NC, June, 1991.

In: Fuzzy Logic
Editor: Dinko Vukadinovic

ISBN: 978-1-62417-151-2

Chapter 5

PROCESSING FUZZY LOGIC BY MOLECULES

Pier Luigi Gentili [*]
Dipartimento di Chimica, Università di Perugia, Perugia, Italy

1. ABSTRACT

Current computers process information based on transistors and electrical signals. The futuristic chemical computers will store, process, and convey information by using molecules, their assemblies, and physical-chemical signals. It is possible to compute by exploiting single molecules or large collections of them. Different kinds of logic can be processed. Since molecules obey the laws of quantum-mechanics, quantum logic can be implemented, as long as decoherent effects are avoided. If the collapse of superimposed or entangled wave-functions is inevitable, molecules can still be used to process either Boolean or discrete multi-valued or fuzzy logic. The conditions favourable to chemically process the infinite-valued fuzzy logic are presented in this text and few examples of its chemical implementation are reported. Fuzzy logic is particularly important for the development of artificial intelligence because it models pretty well human decision making. This property is due to the structural analogies existing between fuzzy logic systems and human nervous system.

[*] pierluigi.gentili@unipg.it.

2. Introduction

Current electronic computers are based on the so-called Von Neumann architecture [1] (see Figure 1) proposed for the first time in the mid-forties. They are constituted by four main elements: (1) a memory storing information; (2) a central processing unit (CPU) wherein information is processed (it consists of two subunits: the arithmetic logic unit, which performs arithmetic and logical operations, and the control unit, which extracts instructions from memory, decoding and executing them); (3) an information exchanger (IE) allowing transfer of information in and out of the computer (it consists of screen, keyboards, etc.); (4) a data and instructions bus (playing as a communication channel) binding the other three components of computer by conveying information among them. Electronic computers are general purpose computing machines because the instructions to make computation are stored into the memory as sequences of bits. Information is encoded in electrical signals, and transistors are the basic switching elements of CPU. One of the main tasks of information technology (IT) is that of devising always more powerful computer, capable of processing larger amounts of information at increasingly higher speed, lower power, volume and price. The ultimate goal is that of making our computing machines able to face as many problems of computational complexity as possible. We need to face "non deterministic polynomial" (NP) and pattern recognition problems efficiently and in reasonable lapses of time. The strategy that IT is pursuing consists of a steady shrinking of the transistors' dimensions by following a top-down approach, through photolithography and related techniques. The pace of the improvements is described by the law formulated by Gordon Moore (co-founder of Intel) stating that the number of transistors on a chip doubles every two years [2].

The race towards always smaller dimensions of elementary computing elements is now approaching some fundamental limits because transistors are made of a few atoms. At this level, a few relevant technological problems arise, such as current leakage (due to quantum-mechanical tunnelling events) and heat dissipation.

To overcome these limitations and pursue improvements in the computational performances, new strategies are required. Plausible and tentative solutions have been proposed by natural computing. It is worthwhile noticing that an expanded computational power would result useful also to deeply understand the intimate secrets of natural complex systems and control their dynamics.

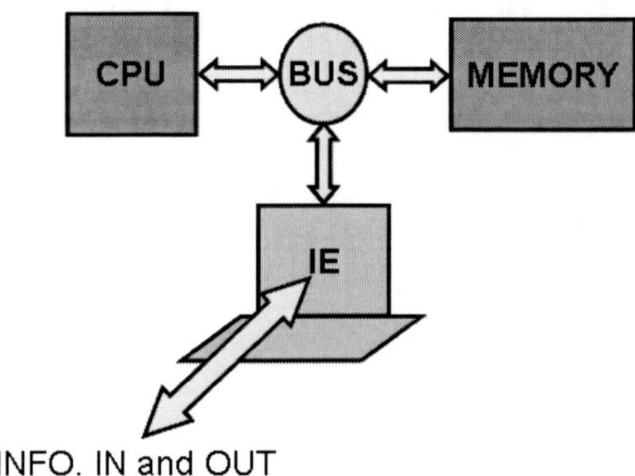

Figure 1. Schematic structure of current computers rooted to Von Neumann architecture.

3. NATURAL COMPUTING

Natural computing is an interdisciplinary scientific discipline based on the rationale that every natural transformation is a kind of computation. It is basically aimed at dealing with complexity (both computational and natural one). It involves three different research lines [3] as they were three prongs of a pitchfork for handling the haystack of complexity. The first research line consists of formulating new algorithms inspired by natural phenomena such as the behaviours of neural networks, immune system, biological evolution, cell, swarms and flocks. These new algorithms are implemented as software on current computers. The second research line regards the development of models to describe natural complex systems, such as cellular automata, fractal geometry, Lindenmayer systems and system biology. The third branch of natural computing is focused on finding new natural materials for solving problems and performing computation. Scientific community reserves great expectations on this latter research line for coping with problems of computational complexity and opening new frontiers in the field of artificial intelligence. In fact, by exploiting atoms and molecules, it is possible to formulate new algorithms and devise brand new computer's architectures.

4. Computing with Atoms and Molecules

4.1. Quantum Computing

If single atoms and molecules are used as computing elements, a paradigm shift in logic is required: classical information must be replaced by quantum information. In other words, the bit (the elementary unit of classical information) must be substituted by the qubit (the elementary unit of quantum information) [4]. A qubit, $|\Psi\rangle$, is a quantum system having two accessible states, labelled $|0\rangle$ and $|1\rangle$, and it can exist as superposition of them. It is represented as a linear combination of the two states:

$$|\Psi\rangle = a|0\rangle + b|1\rangle \tag{1}$$

In equation (1), a and b are two complex numbers verifying a normalization convention:

$$|a|^2 + |b|^2 = 1 \tag{2}$$

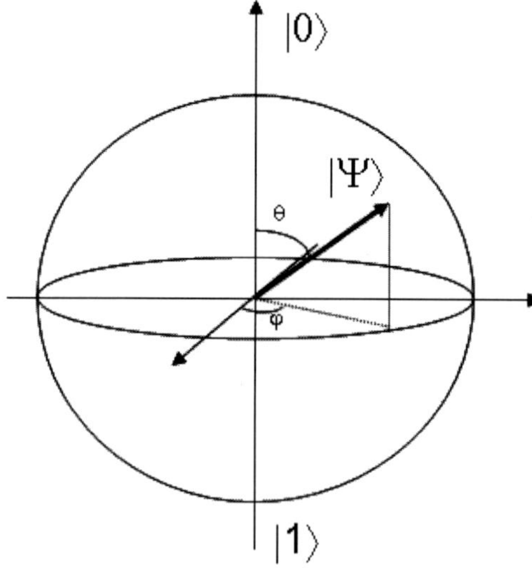

Figure 2. Representation of the qubit $|\Psi\rangle$ in the Bloch sphere.

$|\Psi\rangle$ is a unit vector in a two dimensional Hilbert space. The states $|0\rangle$ and $|1\rangle$ are the computational basis states and form an orthonormal basis for this vector space. The state of the qubit can be also described by equation (3):

$$|\Psi\rangle = \cos(\theta/2)|0\rangle + e^{i\varphi}\sin(\theta/2)|1\rangle \tag{3}$$

wherein θ and φ define a point on a three dimensional sphere, called the Bloch sphere [5] (see Figure 2). Represented on such a sphere, a bit could only lie on one of its poles, representing the pure $|0\rangle$ or $|1\rangle$ states.

Analogously to classical information, also quantum information can be stored, processed and conveyed. Therefore, it is possible to contrive quantum computers consisting of memory, logic gates and wires as their classical counterparts [6]. The quantum logic gates are based on quantum-mechanical phenomena, and are unitary operators, represented by matrices, preserving the norm of quantum states. A quantum logical operation is visualized as rotation of the unit vector in the Bloch sphere. This rotation is always reversible. The result of quantum operation is conveyed not through physical wires, but by travelling particles, such as photons. A quantum computer is drawing much attention not because it can imitate "classical" computation, but because it can do much more. Its computational power relies on the superposition of states and their interference [7]. If a quantum system consists of n unmeasured qubits, it can be in an arbitrary superposition of up to 2^n different states simultaneously, differently from a classical computer that can only be in one of the 2^n states at any one time [8]. The superposition can involve also the quantum states of physically separated particles, if they are entangled [9]. A computation involves all the 2^n superimposed states. In order to take advantage of the intrinsic parallelism, it is necessary to combine it with the interference. The states within a superposition can actually interfere with each other, constructively reinforcing or destructively cancelling each other to form the final definite state. The programmer of a quantum computer must choreograph the calculation in such a way that computational paths leading to a wrong answer will reduce or even destructively interfere, whereas those leading to the right answer will reinforce or constructively interfere.

Working with multiple states can speed up some kinds of computations (for example, the Shor's algorithm for factoring integers [10]) and it makes doable some tasks, such as absolute secrecy of communication between parties (e.g. quantum cryptography [11]) that is impossible classically. The main difficulty in building a quantum computer derives from the pernicious interactions between qubits and the surrounding environment, triggering loss of coherence. For instance, by decoherence the qubit of equation (1) will collapse into $|0\rangle$ or $|1\rangle$ state, with probability a^2 and b^2, respectively. The loss of quantum coherence occurs also whenever quantum states are measured.

So far, different solutions have been proposed to implement quantum computations. An example is offered by a collection of ions tightly held in an electromagnetic trap [12], where each atom stores a qubit in a pair of internal electronic levels. By a laser radiation it is possible to detect the state of each qubit, since only one of the two internal states fluoresces. The charged atoms are coupled by virtue of their mutual Coulomb repulsion. Another example is given by the nuclear spins of atoms belonging to molecules and immersed in a magnetic field. Using nuclear magnetic resonance (NMR) techniques, these spins can be manipulated, initialized and measured [13]. Other proposals consist of some optical systems [14] and solid state systems like array of quantum dots [15]. In these technologies the qubits have usually long relaxation times to prevent quantum superposition from decohering away. With current schemes, high hindrances stand when quantum computers are scaled up, due to the difficulties in creating and maintaining several quantum states.

4.2. "Classical" Computing

Decoherent effects bring about transitions from quantum to classical information [16]. By decoherence, a qubit $|\Psi\rangle$ will collapse into either $|0\rangle$ or $|1\rangle$ state. In this situation, single atoms and molecules can still be exploited as computing elements, not to process quantum but just "classical" binary logic. In the case of three level or n-level quantum systems (with n>3), which have three or n basis states, the decoherence determines the possibility of processing three-valued or multi-valued logics, i.e. crisp logics wherein there are more than two truth values. Under the practical point of view, it is possible to work with single molecules only by exploiting the principles of microscopic techniques reaching atomic resolution, such as the scanning tunnelling and the atomic force microscopies. To process "classical" logic, it is also possible to exploit macroscopic portions of matter [17], consisting of a huge number of computing molecules or atoms (of the order of the Avogadro's number). These large collections of molecules are bulky materials, which can receive macroscopic inputs from the outside and give back output signals to the surrounding macro-environment. Therefore, inputs and outputs become continuous variables. They have chemical or physical nature, such as electrical, optical, thermal, mechanical or magnetic ones. Optical variables are particularly appealing especially if they involve frequencies belonging to the visible region. As input, light can be focused on very small areas and, as output, it has the peculiarity of being easily caught by a human observer, bridging the gap between two worlds: the molecular one with that of the human observer. Moreover, optical signals can be easily conveyed by optical fibres.

Quite often, the relation establishing between input and output variables is well described by the logistic function [equation (4)]:

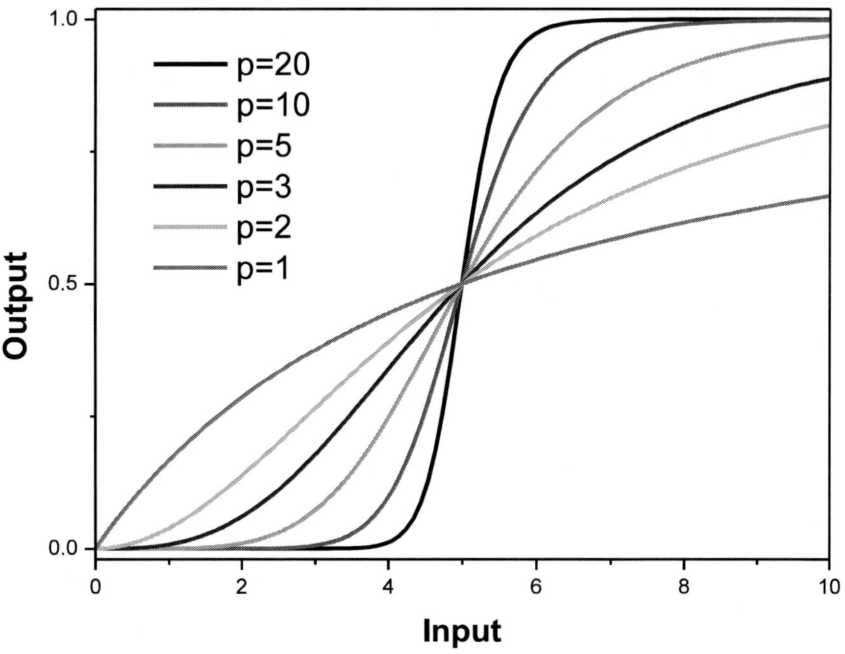

Figure 3. Input-Output relation described by the logistic function (see equation 4) and plotted for different values of the exponential term p.

$$y = \frac{A_1 - A_2}{1 + \left(\frac{x}{x_0}\right)^p} + A_2 \tag{4}$$

When the values of the constants A_1, A_2 and x_0 are fixed, the logistic function assumes different shapes depending on the values of the exponential p (see Figure 3). It has two limiting shapes: a truly sigmoid (when p is large) and an hyperbolic one (when p is small).

If the relation has sigmoid shape with a steep slope around the inflexion point, it is suitable to process Boolean logic. In case the relation is characterized by two or more steps, it is suitable to process three- or multi-valued logic, respectively. On the other hand, if the relation has hyperbolic shape, it is not useful for crisp logics, but for an infinite-valued logic, that is for fuzzy logic. This distinction is also valid in current information technology. In order to process binary logic, the electronic circuits handle sigmoid electrical signals, which vary steeply. On the other hand, the best accomplishments of fuzzy inference engines have been achieved so far by analogue electronic circuits, based on signals varying smoothly, in linear or hyperbolic manner [18].

5. Importance of Fuzzy Logic

The scientists working on artificial intelligence are trying to develop human level machine intelligence [19]. Humans have many remarkable mental capabilities. Two of them stand out. First, the capability to perform a wide variety of physical tasks without any measurement and any mathematical computation. Secondly, the ability to make rational decisions and mental tasks based upon partiality of truth, possibility, and uncertain, imprecise, incomplete information, gathered through perceptions. It is mandatory to implement these two noticeable mental activities in artificial matter, in order to give rise to human level machine intelligence. This tough task is beyond of reach by using just Boolean logic, in agreement with Zadeh's prediction [19]. What is needed is to compute with words. The logic which allows to compute with words is fuzzy logic. In fact, fuzzy logic is a precise logic of imprecision and approximate reasoning. It involves graduated granulation of linguistic variables [19] and if-then rules. So far, fuzzy logic has been implemented either in digital electronic computers mainly by means of specific software or by new hardware underlain on analog electronic circuits [20].

To open new vistas, it is compelling to process fuzzy logic by exploiting other natural materials and other kinds of signals. In the next paragraph, the first examples of chemical implementation of the fundamental fuzzy logic operators are described.

6. Examples of Chemical Implementation of Fuzzy Logic

A first example of a chemical fuzzy logic inference engine is offered by the photophysical behaviour of carbonyl and nitrogen-heterocyclic compounds (see Figure 4) exhibiting proximity effects [21-22]. By exciting these molecules with UV radiation, a qubit is produced, defined by equation (5):

$$|\Psi\rangle = a|\psi_{\pi,\pi^*}\rangle + b|\psi_{n,\pi^*}\rangle \tag{5}$$

The two wave-functions of the linear combination, $|\psi_{\pi,\pi^*}\rangle$ and $|\psi_{n,\pi^*}\rangle$, are relative to the two electronic excited states, (π,π^*) and (n,π^*), primarily associated with the C=O and C=N chemical groups (see left part of Figure 4). Due to the interactions with the surrounding microenvironment, $|\Psi\rangle$ rapidly (in a few nanoseconds after photo-excitation) collapses into the (π,π^*) or (n,π^*) states, with probability a^2 and b^2, respectively. If it collapses into $|\psi_{\pi,\pi^*}\rangle$, the molecule can emit light, whereas if it collapses into $|\psi_{n,\pi^*}\rangle$, the molecule is dark, since it thermally relaxes to the ground state (indicated as (0) in Figure 4) bypassing the (π,π^*) state. The values of the a and b coefficients depend on the strength of the coupling between the (π,π^*) and (n,π^*) states [23]. The extent of their coupling is inversely proportional to the energy gap between them. The wider the energy gap between the (π,π^*) and (n,π^*) states, the weaker the coupling between them. When the two electronic states are weakly coupled, the a coefficient of equation (5) can be large. Therefore, the probability a^2 that the molecular excited state will collapse into $|\psi_{\pi,\pi^*}\rangle$ will be high. If we consider a solution containing a huge number of molecules of either an aromatic carbonyl or an heterocyclic compound, a large a coefficient will mean a high fluorescence quantum yield (Φ_F) for the macroscopic sample. It is possible to control the strength of the coupling and hence the values of the a and b coefficients through macroscopic physicochemical parameters, such as the temperature (T) and the hydrogen bond donating (HBD) ability of solvent [22]. In general, high T favours the coupling, whereas a high HBD ability of the solvent reduces the coupling for an energetic arrangement of the excited states such as that depicted in Figure 4 (where the (π,π^*) state lies below the (n,π^*) state) [22]. Therefore, Φ_F grows up by cooling (i.e. reducing T) and solubilising the compound into a solvent having high HBD ability, as shown for 6(5H)-Phenanthridinone in the right part of Figure 4. It is noteworthy that Φ_F changes smoothly, in hyperbolic rather than sigmoid manner: its dependence on T and HBD can be exploited to process fuzzy logic. Fuzzy logic systems (FLSs) have been built by both the Mamdani's and Sugeno's methods [24]. They involved T and HBD of solvent as inputs, Φ_F as output and UV radiation as power supply. The fuzzy rules formulated in the FLSs had multiple antecedents connected just through the AND operator. The reset time of the molecular system was a few nanoseconds.

Another example of molecular implementation of the AND fuzzy operator is based on the photo-response of the natural amino acid tryptophan, either isolated or bound to a protein [25]. By absorbing UV-A photons, tryptophan molecules jump to their first electronic excited state (S_1) (see the scheme on the gray patch of Figure 5). From this unstable state, they relax in a few nanoseconds by following different pathways. They can decay either emitting light, or squandering the stored electromagnetic energy in heat, or participating in an electron/proton transfer reaction, or transferring their excess energy to another species, such as flindersine (FL). These relaxation routes are in kinetic competition among them. The faster the route, the higher the probability of occurring and hence its quantum yield. It is possible to control the rates of a few of these processes through specific physical and chemical inputs. For instance, by increasing T we favour the reactive thermally-activated paths, whereas by adding many molecules of flindersine, we favour the energy transfer process. It derives that a macroscopic sample containing a large number of tryptophan molecules exhibits a fluorescence quantum yield Φ_F which increases by reducing T and limiting the content of the quencher flindersine. Since the dependence of Φ_F on T and the flindersine-over-tryptophan molar ratio (n_{FL}/n_{Try}) is smooth, it is ideal to process the infinite-valued fuzzy logic. Fuzzy logic systems (FLSs) have been accomplished by both the Mamdani's and Sugeno's method [25]. The temperature and the flindersine-over-tryptophan molar ratio were the inputs, Φ_F of tryptophan was the optical output, and UV radiation was the power supply. These FLSs were based on fuzzy rules with multiple antecedents connected just through the fuzzy AND operator. An example of partition of the input and output variables in fuzzy sets, according to the Mamdani's method, is illustrated in Figure 5. The reset time of this fuzzy AND operator was of the order of a few nanoseconds, as the previously one.

Fuzzy logic systems involving all the fundamental fuzzy operators, AND, NOT, OR, can be implemented by the chromogenism of the spiro-oxazine (SpO) shown in Figure 6A. UV irradiation brings about its spiro C-O bond breakage producing a merocyanine (MC) which gives rise to an absorption band in the visible region of the spectrum, centred at 611 nm (see Figure 6B) [26]. Its solution becomes coloured in blue (all the transmittance spectra, shown in Figure 6, have been recorded in acetonitrile with SpO solved in concentration of the order of 10^{-5} mol dm^{-3}). MC reverts back to SpO thermally, in few hundreds of seconds (at room temperature) and the solution becomes uncoloured, again. When the UV irradiation is carried out in the presence of equimolar amounts of ions respect to the moles of SpO, different transmittance spectra are recorded at the photo-stationary state. In the presence of protons (H^+) and aluminium cations (Al^{3+}), an absorption band centred at 485 nm appears and the solution becomes orange. In the presence of cupric cations (Cu^{2+}), a band centred at 423 nm comes out and the solution becomes yellow (see Figure 6C). The initial molecular state, SpO, can be restored in a few seconds by addition of specific chemicals neutralizing the effect of the inputs (see reference [26] for more details). If the cations, H^+, Al^{3+}, Cu^{2+}, are added in amounts less than the equimolar one respect to SpO, spectra exhibiting combinations of the coloured bands (those centred at 611, 485 and 423 nm) with different height can be recorded. The solution changes its colour in chameleonic way, from blue to cyan, to grey up to orange by progressively increasing the amount of H^+ or/and Al^{3+}. It goes from blue to green, to different tones of yellow by injecting growing amount of Cu^{2+}. Other colours can be obtained by adding different combinations of $H^+ + Cu^{2+}$ or $Al^{3+} + Cu^{2+}$.

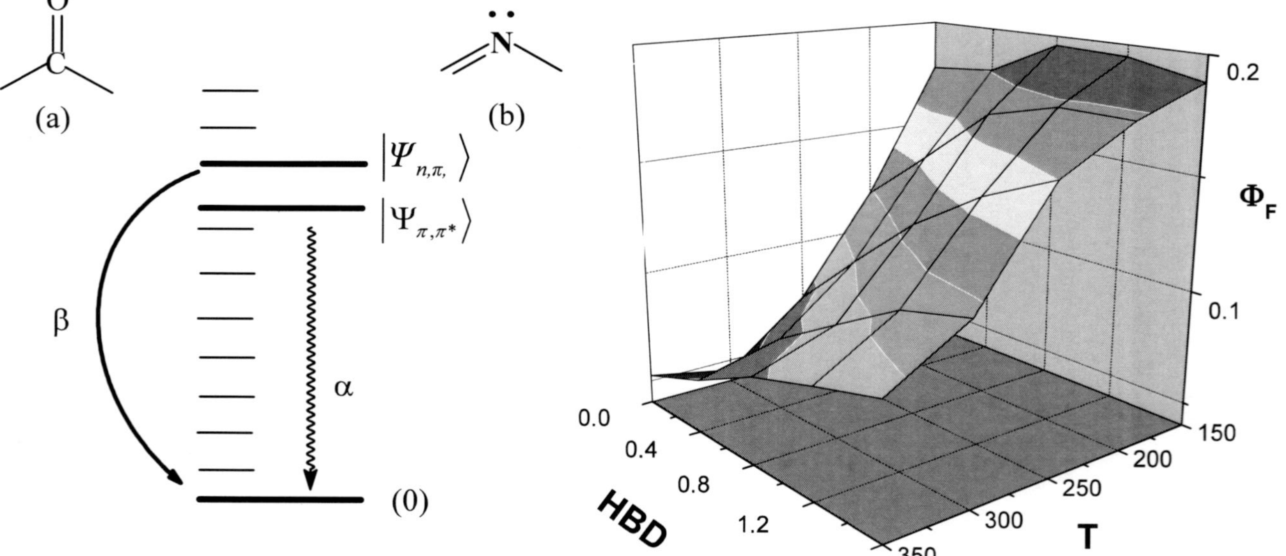

Figure 4. On the left, scheme representing the relaxation dynamics for (a) an aromatic carbonyl or (b) a nitrogen heterocyclic compound after photo-excitation: the arrow marked as α represents the fluorescent decay, whereas that marked as β is the dark relaxation to the electronic ground state (0). On the right, three-dimensional plot showing the dependence of the fluorescence quantum yield (Φ_F) of 6(5H)-Phenathridinone on temperature (T) and Hydrogen Bond Donation (HBD) power of the solvent.

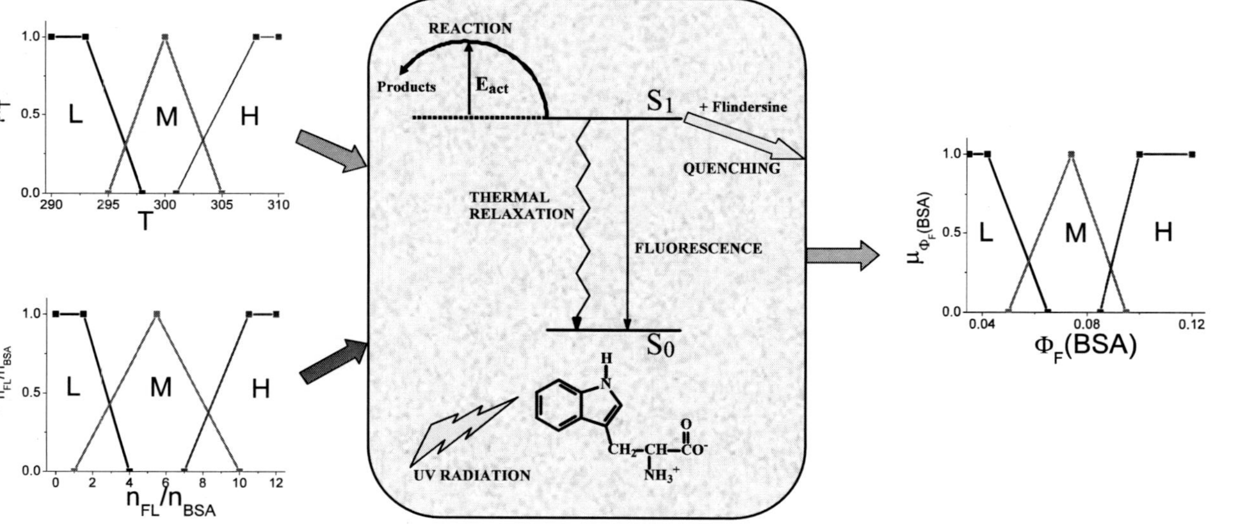

Figure 5. Representation of a fuzzy logic system based on the tryptophan's photo-behaviour (which is sketched inside the gray patch) and built by the Mamdani's method. Tryptophan is bounded to Bovine Serum Albumin (BSA). T and the molar ration n_{FL}/n_{BSA} are the inputs, Φ_F is the output. All the variables are partitioned in three fuzzy sets (labelled Low (L), Medium (M), High (H), respectively).

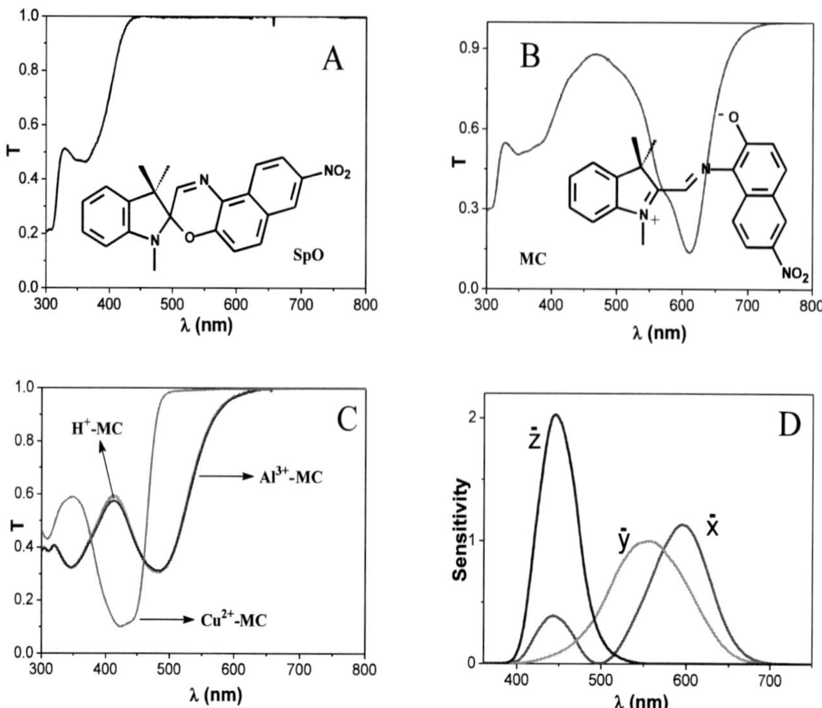

Figure 6. Transmittance spectra ($T=I_T/I_0$, where I_T and I_0 are the intensities of the transmitted and incident radiations, respectively), and molecular structures of SpO (A) and MC (B). (C) Transmittance spectra of H^+-MC (green trace), Al^{3+}-MC (blue trace), and Cu^{2+}-MC (magenta trace). (D) Colour-matching functions defined by CIE in 1964..

It results that by using UV radiation as power supply and analog injection of H^+, Al^{3+}, Cu^{2+} as chemical inputs, the solution can acquire an "infinite" number of colours (reminding that two colours are different when just one of the three R, G, B colour coordinates assumes a different value) because there are plenty of photo-stationary states which are achievable. Therefore, the chromogenic properties of SpO can be exploited to process fuzzy logic, having the transmitted intensities of light at the different wavelengths of the visible region as output. For the granulation of the output, the fuzzy nature of human colour perception [17, 27] has been imitated. Human beings have three types of cones whereby they distinguish colours. The absorption spectra of the three sorts of cones can be assumed to be three fuzzy sets. When visible radiation, consisting of a mixture of several wavelengths having different intensities, hits the retina of our eyes, it activates the three types of cones in a specific proportion which is function of the spectral composition of the radiation itself and the spectral sensitivity of the cones. Radiations of different colours will have different values of membership functions in the three fuzzy sets. The combination of these values are transduced into the perception of specific colours inside our brain. For the FLSs based on the chromogenism of SpO, the visible spectral region has been partitioned in the three colour-matching functions, $\bar{x}$, $\bar{y}$, $\bar{z}$, whereby the CIE (Commision Internationale de l'Éclairage) standardized the sensitivity of human eye in 1964 (see Figure 6D). The defuzzification procedures consisted in transforming the transmittance spectra ($T(\lambda)$) into the X, Y, Z tristimulus values, according to the following three integrals (equation (6)):

$$
\begin{aligned}
X &= \frac{1}{k}\int_{360}^{800} D(\lambda)T(\lambda)\bar{x}(\lambda)d\lambda \\
Y &= \frac{1}{k}\int_{360}^{800} D(\lambda)T(\lambda)\bar{y}(\lambda)d\lambda \\
Z &= \frac{1}{k}\int_{360}^{800} D(\lambda)T(\lambda)\bar{z}(\lambda)d\lambda
\end{aligned}
\tag{6}
$$

wherein $D(\lambda)$ is the energy distribution of the CIE normalized illuminant D65 (which closely matches that of the sky daylight), and k is a normalization factor defined in such a way that an object with a uniform transmittance $T(\lambda)=1$ gives a luminance component Y=1 [26]. The X, Y, Z tristimulus values can be transformed into the R, G, B coordinates by a linear transformation.

Two FLSs have been built, differing in the combination of the inputs: one has been based on H^+ and Al^{3+}, whereas the other on H^+ and Cu^{2+} as chemical inputs. The former involved all the fundamental fuzzy operators, whereas the latter just AND and NOT.

There exist also other chemical reactions which allow fuzzy logic to be chemically processed. An example is the Belousov-Zhabotinsky (BZ) reaction, that is the catalytic oxidative bromination of an organic substrate, such as malonic acid, by bromate in acidic medium [28]. When the BZ reaction is in oscillatory regime, it is formally similar to the dynamical behaviour of biological excitable/oscillatory cells (i.e., neurons), which spontaneously depolarize their axon hillock and fire actions potentials, often at regular rate. Although such natural oscillators have their own internal rhythm, external stimuli can alter their timing. In pacemaker cells, for example, information about a stimulus is encoded by changes in the timing of individual action potentials, and it is used to rule proprioception and motor coordination for running, swimming, and flying [29]. The same is true for the BZ reaction. Perturbation of the BZ reaction by different species shifts the phase of oscillations. The addition of Br^- leads to a delay (it plays as an inhibitor), while removal of Br^-, by injection of silver ions (playing like activators), leads to an advance [30]. In reference [31] there are the experimental evidences demonstrating that smooth variations of variables such as the volume and the instant of injection of activator and inhibitor solutions determines smooth variations of the period of oscillations. Therefore, their relations are suitable to process fuzzy logic. Some FLSs have been built based on rules involving all the fundamental fuzzy logic operators. Although the detailed workings of the oscillating BZ reaction are quite different from those of pacemaker cells, the similarities in their dynamical behaviour suggest that a pacemaker cell could process fuzzy logic in much the same fashion as the BZ reaction does.

Finally, it is worthwhile mentioning two other chemical systems suitable to process fuzzy logic. The first is the biochemical reaction network controlling the glycolysis/gluoconeogenesis functions [32]. Here, fructose-6-phosphate (F6P) is interconverted between its two bisphosphate forms by specific kinases and phosphatases. The enzymes in this kinetic mechanism are under the allosteric control of many of the chemical signals of cellular energy status such as cyclic-adenosine-monophosphate (cAMP) and citrate. The dependence of the concentration of F6P on those of cAMP and citrate, gives rise to a 3D surface showing a not abrupt transition from low to high values, such as that of Figure 4. The profile of the 3D surface has a smooth hyperbolic shape and not a steep sigmoidal response: it is suited to process fuzzy logic.

A final chemical example is the DNA hybridisation wherein two single-stranded DNA molecules (oligonucleotides) bind to form a double stranded DNA duplex. At room temperature, the hybridisation reaction is not a two-state, all or none process, but it is inherently fuzzy because it is a continuum of outcomes [33]. The pairs of oligonucleotides formed inside a test tube cannot be divided into distinct sets of hybridised and unhybridised species, but each molecule would have a degree of membership in both.

CONCLUSION

The research field of natural computing is paving new paths for facing both the computational and the natural complexity. A promising route consists in finding alternative natural materials to compute. If computation is performed by means of atoms and molecules, different kinds of logic can be processed. As long as decoherent effects are avoided, it is possible to compute exploiting the laws and the dynamics of the quantum world. Whenever the collapse of quantum wave-functions occurs too fast and is unavoidable, it is still feasible to use atoms and molecules as computational tools. However, it is allowed to process just logics which are based on the laws and the dynamics of classical physics. Therefore, binary or multi-valued crisp logics can be processed by working with single molecules, whereas both crisp and fuzzy logics can be implemented by using large collections of elementary computing elements. Among the "classical" ways of computing, fuzzy logic is drawing much attention because it can contribute to the development of human level machine intelligence. It allows to compute with words, i.e. it handles, in accurate manner, uncertain and imprecise information.

So far, there are a few examples of chemical implementation of fuzzy logic operators: those herein described. Two of them, i.e. the chromogenic SpO and the oscillating BZ reaction, are particularly attractive because they allow all the fundamental fuzzy logic operators to be implemented. The fuzzy logic systems based on the oscillating BZ reaction have the advantage to showing a faster reset time: a few tens of seconds rather than hundreds of seconds. However, the FLSs based on the chromogenism of SpO are particularly appealing because they are based on a defuzzification procedure imitating the way humans distinguish colours. The perception of colours, as any other physical perceptive human capability is inherently fuzzy [17]. In fact the structure of the human nervous system is strictly related with the structure of a fuzzy logic system. The ensemble of receptor cells (like the cones into the vision) works as a fuzzifier; the brain acts as if it were the fuzzy inference engine and the effectors' cells behave as defuzzifier. This analogy will surely inspire new ways of implementing fuzzy logic at the molecular level for the next future.

REFERENCES

[1] Burks, A. W.; Goldstine, H. H.; von Neumann, J. *Preliminary discussion of the logical design of an electronic computing instrument.* In Taub, A. H., editor, John von Neumann Collected works, The Macmillan Co., New York, Volume V, 1963, 34-79.

[2] Moore, G. Lithography and the Future of Moore's Law. *Proceedings of SPIE,* May 1995, 2437, 1-8.

[3] de Castro, L. N. Fundamentals of natural computing: an overview, *Phys. Life Rev.* 2007, 4, 1-36.

[4] Schumacher, B. Quantum coding, *Phys. Rev.* A 1995, 51, 2738–2747.

[5] Chruściński, D. Geometric Aspect of Quantum Mechanics and Quantum Entanglement. *J. Phys. Conf. Series* 2006, 39, 9-16.

[6] Feynman, R. P. Simulating Physics with Computers. *Int. J. Theor. Phys.* 1982, 21, 467-488.

[7] Aharonov, D. Quantum Computation- A Review. *Annual Review of Computational Physics,* World Scientific, volume VI, ed. Dietrich Stauffer (1998).

[8] Bennet, C. H.; DiVincenzo, D. P. Quantum information and computation. *Nature.* 2000, 404, 247-255.

[9] Plenio, M. B.; Virmani, S. An introduction to entanglement measures, *Quant. Inf. Comp.* 2007, 7, 1-51.

[10] Shor, P. W. Polynomial time algorithms for prime factorization and discrete algorithms on a quantum computer. *SIAM J. Comput.* 1997, 26, 1484-1509.

[11] Bennett, C, H.; Bessette, F.; Brassard, G.; Salvail, L.; Smolin, J. Experimental Quantum Cryptography. *J. Cryptology* 1992, 5, 3-28.

[12] (a) Cirac, J. I.; Zoller, P. Quantum Computation with Cold Trapped Ions. Phys. Rev. Lett., 1995, 74, 4091-4094. (b) Monroe, C.; Meekhof, D. M.; King, B. E.; Itano, W. M.; Wineland, D. J. Demonstration of a Fundamental Quantum Logic Gate. *Phys. Rev. Lett.* 1995, 75, 4714–4717.

[13] Cory, D. G.; Fahmy, A. F.; Havel, T. F. Nuclear magnetic resonance: an experimentally accessible paradigm for quantum computing. *Proceedings of PhysComp* 1996, 87-91.

[14] Knill, E.; Laflamme, R.; Milburn, G. J. A scheme for efficient quantum computation with linear optics. *Nature* 2001, 409, 46-52.

[15] DiVincenzo, D. P.; Bacon, D.; Kempe, J.; Burkard, G.; Whaley, K. B. Universal quantum computation with the exchange interaction. *Nature* 2000, 408, 339-342.

[16] Zurek, W. H. Decoherence and the Transition from Quantum to Classical. *Phys. Today* 1991, 44, 36 –44.

[17] Gentili, P. L. Molecular Processors: From Qubits to Fuzzy Logic. *ChemPhysChem* 2011, 12, 739-745, and references therein.

[18] Zadeh, L. A. Fuzzy Logic. *Computer* 1988, 83-93.

[19] Zadeh, L. A. Toward Human Level machine Intelligence- Is It Achievable? The Need for a Paradigm Shift. *IEEE Computational Intelligence Magazine* 2008, 3, 11-22.

[20] (a) Yamakawa, T.; Electronic Circuits dedicated to fuzzy logic controller. Scientia Iranica, 2011, 18 (3), 528-538. (b) Yamakawa, T.; Miki, T.; Ueno, F.; Basic Fuzzy Logic Circuit Formed by Using p-MOS Current Mirror Circuits, Electronics and Comm. in Japan, 1986, 68, 2, 1-9. (c) Yamakawa, T.; Miki, T.; The Current Mode Fuzzy Logic Integrated Circuits Fabricated by the Standard CMOS Process. IEEE Trans. On Computers, 1986, 35, 2, 161-167. (d) Guo, S.; Peters, L.; Surmann, H.; Design and Application of an Analog Fuzzy Logic Controller. IEEE Trans. *On Fuzzy Systems* 1996, 4, 429-438.

[21] (a) Lim, E. C.; Proximity Effect in Molecular Photophysics: Dynamical Consequences of Pseudo-Jahn-Teller Interaction. *J. Phys. Chem.* 1986, 90, 6770-6777.

[22] Gentili, P. L.; Ortica, F.; Romani, A.; Favaro, G. Effects of Proximity on the Relaxation Dynamics of Flindersine and 6(5H)-Phenanthridinone. *J. Phys. Chem.* A 2007, 111, 193-200.

[23] Siebrand, W.; Zgierski, M. Z. Radiationless decay of vibronically coupled electronic states. II. Non-Condon effects of nontotally symmetric accepting modes. *J. Chem. Phys.* 1980, 72, 1641-1646.

[24] Gentili, P. L. Boolean and fuzzy logic implemented at the molecular level. *Chem. Phys.* 2007, 336, 64-73.

[25] Gentili, P. L. Boolean and Fuzzy Logic Gates Based on the Interaction of Flindersine with Bovine Serum Albumin and Tryptophan. *J. Phys. Chem.* A 2008, 112, 11992-11997.

[26] Gentili, P. L. The fundamental Fuzzy logic operators and some complex Boolean logic circuits implemented by the chromogenism of a spirooxazine. *Phys. Chem. Chem. Phys.* 2011, 13, 20335–20344.

[27] Gentili, P. L. Fuzzy Logic in Molecular Computing. *Expert Commentary in "Fuzzy Logic: Theory, Programming and*

Applications", Editor R. E. Vargas, Nova Science Publishers, Inc. 2009, pag 1-10.

[28] Epstein, I. R.; Pojman, J. A. *An Introduction to Nonlinear Chemical Dynamics.* Oxford University Press: New York. 1998.

[29] Winfree, A. T. Phase-Control of Neural Pacemakers. *Science* 1977, 197, 761-763.

[30] Ruoff, P. Phase Response Relationships of the Closed Bromide-Perturbed Belousov-Zhabotinsky Reaction – Evidence of Bromide Control of the Free Oscillating State without use of a Bromide-Detecting Device. *J. Phys. Chem.* 1984, 88, 2851-2857.

[31] Gentili, P. L.; Horvath, V.; Vanag, V. K.; Epstein, I. R. Belousov-Zhabotinsky "chemical neuron" as a binary and fuzzy logic processor. *Int. J. Unconventional Computing* 2012, 8, 177-192.

[32] Arkin, A.; Ross, J. Computational Functions in Biochemical Reaction Networks. Biophys. J. 1994, 67, 560-578.

[33] Deaton, R.; Garzon, M. Fuzzy logic with biomolecules. *Soft Computing* 2001, 5, 2-9.

INDEX

C

D

E

F

G

H

I

J

K

L

M

N

O

P

Q

R

S

T

U

V

W

Y